design with reinforced plastics

a guide for engineers and designers

Rayner M Mayer

The Design Council

Published in the
United Kingdom in 1993 by
The Design Council
28 Haymarket
London SW1Y 4SU

Printed and bound in the
United Kingdom by
Bourne Press Ltd
Bournemouth

British Library in Publication Data
A catalogue record for this book is available from the British Library.

ISBN 0 85072 294 2

Preface

Adding bundles of strong fibres to weak matrices produces a range of materials which are attractive to designers and engineers. Wood is a prime natural example of such a material. The development of suitable synthetic plastics and fibres, especially since 1945, has led to the growth of the fibre reinforced plastics industry.

As experience and confidence in these materials has grown, their use has extended from decorative and functional applications to structural ones. The state of the art has developed to the point where it is often hard to tell whether components are made from reinforced plastics or from metals!

This guide is aimed at readers who are acquainted with the design process and who now wish to design in fibre reinforced plastics rather than other types of material. Although there are many reasons for considering such materials, the designer needs to be aware of what is and is not possible with today's materials and production techniques.

The guide is based on the experience of the authors over the past 10 years, supplemented by what appears in the literature and the results of discussions with designers and engineers in the industry. It discusses all aspects of the design process, starting with the design brief and progressing through the various stages of design to manufacture and testing, based on the new design code BS 7000.

The discussion is always taken from first principles to a level at which existing sources of information and reference texts can be consulted and structural formulae used.

The book covers the impact of the single European market on product design. This includes not only the global directives on product liability and safety, but also the 'new approach' directives and the development of European standards, product marks and test methods.

Comments from users of this guide would be welcomed to enable future editions to be improved.

For additional information on design data, a companion volume entitled *Design Data for Reinforced Plastics* has been published by Chapman & Hall, London.

Rayner M Mayer
Yateley
1992

This guide was prepared in collaboration with:
Alastair Johnson
Graham Sims
of the National Physical Laboratory, Teddington, UK, with additional contributions from:
Allan Court
James Hendry
Walter Lee
Ian Montgomery
of the National Engineering Laboratory, East Kilbride, UK.

It could not have been written and produced without the help of many people throughout the industry. In particular I would like to thank the following:

David Hughes, Optimat, for mooting the concept and providing guidance throughout the preparation of the text

Ken Wallace of Cambridge University, and Peter Hills of the Design Council, on design

Christer Erickson, Det Norske Veritas, and Robin Walley, Lloyds Register, on structural design and verification

Michael Litton, Culzean Fabrics, and Ken Forsdyke, Fortech, on material

Graham Walker, Health and Safety Executive, on safety

Mark Scudamore, BP Chemicals, and Ted Osborn, London Underground, on fire performance;

Graham Chalmers, Mike Gilbert, John Stratford, Geoffrey Strawbridge, BSI on standards

Gill Money, Design Council, and Alan Thompson for comments and assistance with the manuscript.

The guide could not have been researched and written without funding from the Materials Matter Programme of the Department of Trade and Industry.

Fibreglas is a registered trade mark of Owens Corning Fibreglass. R glass is a registered trade name of Vetrotex International. DeraKane is a registered trade mark of Dow Chemicals, Kevlar is a registered trade mark of du Pont Nemours.

The first three chapters of the book follow the process widely used in design, and now enshrined in BS 7000, Guide to Managing Product Design, in which a design is conceived and refined in iterative stages of conceptual, embodiment and detailed design.

A prime concern has been to indicate where fibre reinforced plastics differ from other materials and how these aspects need to be considered in the design process. Flow diagrams are provided as an aid to check the progress of the design.

The impact of the single European Market on design is discussed in chapter 4 with reference to the major harmonization directives, standards, product liability and safety; in particular, fire safety is considered in some detail.

More specific information is presented in three further chapters dealing with materials, processing and testing. Four products involving innovative applications are outlined in the final chapter to illustrate the advantages of using such materials

Within each chapter, tables are used to summarize information and provide check-lists while more detailed information has been separated from the text and put into boxes for clarity. Comprehensive reference information is provided at the end of each chapter to enable each aspect to be considered in more detail.

How to use the book

There are a number of different levels at which the designer can seek information about reinforced plastics.

Those not well versed in such materials should begin with Chapter 1 (on conceptual design) to establish whether the potential advantages of using reinforced plastics outweigh their disadvantages. An example is provided to illustrate how the information provided later in the book can be used.

Flow diagrams are provided at the beginning of Chapter 2 which set out a systematic approach to the design process. The steps are described in this chapter and discussed further in Chapter 3, on detail design.

Cross-references are provided throughout the book so that the design aspects and specific information in Chapters 4 to 7 are closely linked.

The index, list of contents and lists of tables and boxes should also be consulted for source information.

A glossary is provided to define the technical terms used within the text and the industry. The terminology is self-consistent with that adopted by the American ASM International Handbook Committee as being the most authoritative available.

Contents

Conceptual design considerations

chapter 1

summary

This chapter considers the properties that make fibre reinforced plastics attractive to the designer, and outlines their advantages and disadvantages. It goes on to discuss the development of design concepts, working from a design brief and performance specification, with descriptions of specific examples.

Introduction

Over many centuries, man has developed the skills to design and manufacture with such diverse materials as stone, wood, ceramics and metals. Polymeric materials became readily available in the 1950s, as a by-product of the refining of oil. They are now used for a great variety of products, even though there are limits to their load-bearing capability and stiffness.

However, adding reinforcement to polymers in the form of fibres which are inherently stiff and strong has given rise to a unique family of materials which combine the properties of the reinforcement with the processing ability of the plastic. This guide is concerned with the use of these materials by the designer.

The layout of the book follows that of modern design methods which start with a design brief or product specification and establish a conceptual design (or designs) based on particular materials or processes. This chapter begins by considering the properties and advantages of fibre reinforced plastics and how these have been successfully incorporated into a number of specific products.

1.1 Properties

The properties that designers generally find attractive in fibre reinforced plastics include those set out in *Table 1.1*.

Table **1.1**

Attractive properties of the materials

properties
- low density
- high strength
- high stiffness
- corrosion resistance
- wear resistance
- low heat transmission
- good electrical insulation
- low sound transmission

The designer's task is to realize one or more of these properties within a product as required by the market. Considering these properties in more detail:

Low density arises because the density of both the plastic matrix (or resin) and the reinforcement are low compared with that of metals. This translates directly to low mass when used in functional or decorative applications where the structural support is provided by another member, for example in panels used for interior linings of buildings.

High strength and **stiffness** can be achieved by selecting a suitable reinforcing fibre *(Figure 5.1)*. This is particularly attractive when combined with low density to give high values of the specific stiffness and strength. For these reasons such materials have been used in applications like sports goods or in the automotive industry, and for rotating or dynamic components *(Chapter 8)*.

Corrosion resistance is inherent in many types of plastics and translates directly to reinforced plastics provided that the fibres are kept away from the surface itself. For this reason such materials are widely used in pipes and tanks where corrosive liquids or environments are encountered and in marine hulls and structures.

Wear-resistant surfaces can be produced by using either a hard resin, as in brake linings, or an abrasive filler within the gel coat.

Low heat transmission can be useful where good heat insulation is required or if heat is to be localized. For example, a flame attacking a casing made from phenolic resin will only transmit heat to the inside slowly and the casing itself will withstand the heat of the flame. Resins for such applications need to be selected with some care *(Section 4.7).*

Good electrical insulation is also an inherent feature of plastics and the reinforcement will only alter this if the fibres themselves are conductors (like a metal braid).

Low sound transmission arises because plastics have good damping (or attenuation) properties. Reinforced plastic casings are quieter than their metallic equivalents – for example rocker box and timing gear covers on vehicle engines.

Properties such as these can be achieved because of the wide range of fibres and resins available, which can be combined in numerous ways.

1.2 Advantages

As well as having a range of attractive properties, fibre reinforced plastics provide further advantages in terms of economics, manufacture and materials usage *(Table 1.2).*

Table **1.2**

Further advantages with processing

advantages
- textured surfaces
- self colouring
- integration of parts
- economy of scale
- moulding direct to final dimensions
- efficient use of materials
- durability
- lifetime costings attractive

It is instructive to consider how some of these advantages have been used in design:

Textured surfaces are possible because the product will assume the texture of a mould into which the mixture of fibres and resin has been cast or pressed. Textured surfaces can be directly achieved without any finishing, as in medical cabinets, or roughened surfaces to provide non-slip areas on boat decks.

Self-colouring is obtained by adding a suitable dye either to the resin or to the surface layer (if a gel coat is used) as part of the manufacturing process *(Section 6.2).* Furthermore, items can be both self-coloured and textured, for example housings for coin operated telephones or office equipment.

Integration of parts is possible because of the ability to cast or mould complex sections, which has the effect of reducing both manufacturing time and assembly cost. A recent automotive example is the front end of of the Peugeot 405, which incorporates in one moulding some 30 parts which were formerly manufactured separately and joined *(Box 1.1)*.

Similarly, the number of individual parts in the tail fin of the Airbus A310 aeroplane was reduced from over 2,000 in the metal design to under 200 in its composite replacement. The ability to use fibre reinforced plastic efficiently with little waste was also an important consideration.

Economy of scale can occur with both small and large volume production runs. With small volumes relatively simple and cheap tooling can be used which can subsequently be upgraded as volume increases *(Box 1.1; Section 2.6)*.

Box **1.1**

Production of automotive parts

Automotive parts

The technical challenge has been to produce components of a quality and consistency which compete with metallic versions and can be used adjacent to metallic parts. Parts have also to compete in terms of cost, performance and mass saving.

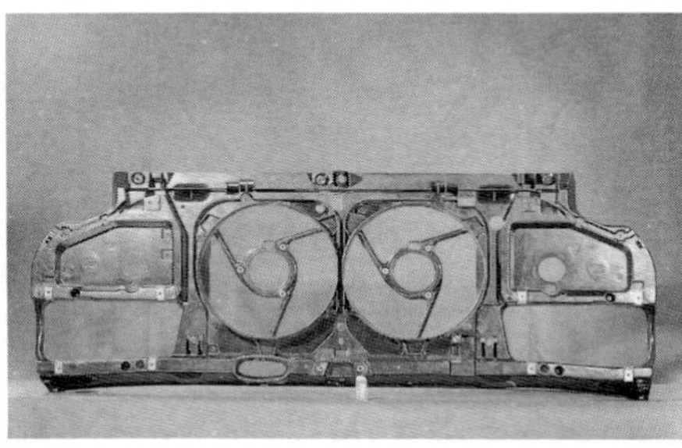

a

b

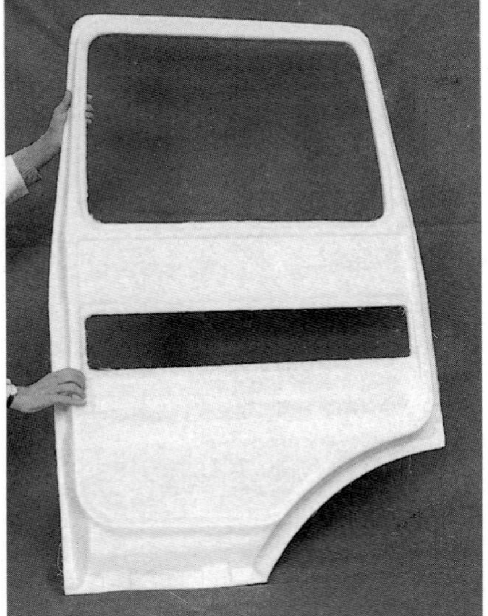

c

Figure **1.1**

Automotive components fabricated by different processes: (a) Front panel, Peugeot 405 by press moulding (b) Tailgate, Citroën AX by compression injection moulding (c) Door, Renault Espace by resin transfer moulding (Buisson)

Fabrication

In processing, attention has been focused on ensuring that the fibre reinforcement is where it is required, that surface quality (as moulded) is good enough to be directly spray painted without any other finishing, and that the component can withstand the high paint stoving temperatures generally associated with metallic parts.

Processes and materials have been developed which have subsequently been used with considerable success in many other applications. These include:

- developing a range of moulding compounds comprising a mixture of catalysed resin, mineral filler and glass fibres of various lengths (such as sheet moulding compounds or SMC) which can be subsequently press moulded *(Figure 1.1) (Section 6.3).*
- maintaining the length and spatial distribution of short fibres in moulding compounds during moulding *(Figure 1.1) (Section 6.9).*
- preforming reinforcement fabrics prior to injecting the resin into a closed mould *(Figure 1.1) (Section 6.4).*

Low volume

Economy of scale is a further feature of working with these materials. Short production runs are possible by using low-cost tooling (often also made from fibre reinforced plastics) and manufacturing direct to final size and colour, using either contact or closed mould techniques *(Sections 6.2, 6.4).* Vehicle manufacturers that have generally incorporated a large proportion of reinforced plastics in their designs include Lotus and Reliant within the UK, and Porsche in Germany. Other examples include vehicles with specialist bodywork such as fire tenders or water carriers.

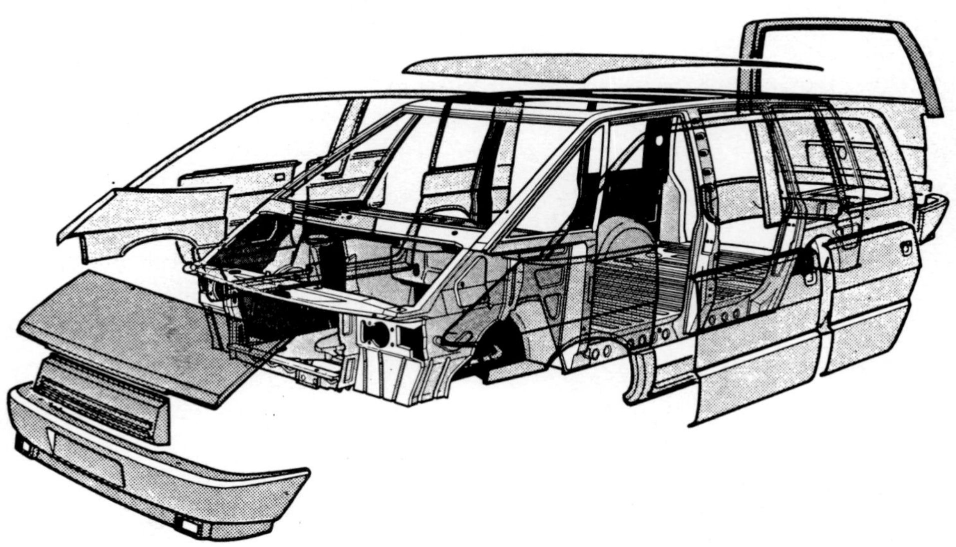

Figure **1.2**

Renault's Espace illustrating variety of panels made from reinforced plastics (Renault Industrie)

Medium volume

The Renault Espace is made in larger quantities, increasing over a period of five years to about 200 a day. All its bodywork is produced using composites: the large parts, such as side panels, roof and rear door, by preforming the fabric reinforcement and moulding by the resin transfer process *(Section 6.4)*, and the horizontal panels by press moulding of SMC *(Section 6.3)*. This vehicle could not have been marketed at an acceptable price at the anticipated sales rate using conventional steel panels formed on steel tooling. ▶

High volume
The production rate is much higher for the bonnet and hatchback door of the Citroën AX and BX models (1,700 a day)*(Figure 1.1)*. Yet this is also possible using either press moulding alone or a mixture of injection and press moulding *(Section 6.3)*. The bonnet is both stiffer and lighter than its metallic counterpart and also incorporates ducts to vent the air flow to the appropriate parts of the engine.

Conclusion
Control over fibre orientation and distribution during processing is the key and has led to a number of innovative processes. This market provides an outlet of sufficient size for developing new production technology for reinforced plastics.

Moulding direct to final dimensions is another notable advantage in that it minimizes material waste and eliminates machining and finishing operations. It is also essential if one wants to use two other advantages already described – namely texturing the surface and self-colouring. Boat hulls and decks are notable examples of these manufacturing techniques and many other marine products are made from these materials for this reason. Examples of two very different products are given in *Boxes 1.2* and *1.3.*

Durability is a feature of many products made from reinforced plastic materials. The first glass fibre reinforced plastic (GRP) components subjected to load were made in 1947 in the form of storage tanks and as yet they have shown little sign of deterioration. Some 600,000 boats have been manufactured according to the Nordic boat standard *(Section 4.4)* and some craft have now been in service for 30 years or more. What has emerged is the importance of correct selection of materials and proper fabrication.

Lifetime cost includes both initial material and manufacturing costs plus those of servicing the product during its lifetime. Materials that are stable will lead to products with lower lifetime costs, even if their initial cost is higher. Reinforced plastics are capable of providing long-life products due particularly to their good corrosion resistance.

1.3 Disadvantages

In order to put the technology into perspective, the attractive properties and advantages of reinforced plastics must be offset with some potential disadvantages *(Table 1.3)*. None of these is insurmountable, but they may well require additional work and therefore extra cost. The question of whether the potential advantages outweigh this extra cost is discussed throughout this guide.

Table **1.3**

Potential disadvantages of using reinforced plastics

potential disadvantages
- properties not established until manufactured
- limited availability of design data
- reinforcement incorrectly located
- lack of codes and standards
- recycling not easy
- poor public acceptance
- fire, smoke and toxicity performance

Box **1.2**

Hammer handles

Hammer handles are a good illustration of the way in which the designer can exploit the varying properties and advantages of reinforced plastic materials.

Three properties are considered of importance: the mass of the handle, its ability to absorb impact energy and the possible shrinkage of the handle within the head. These are rated in a subjective manner in *Table 1.4* for the three principal types of material – wood, steel and glass reinforced plastic (GRP).

From a design viewpoint, it would appear that the GRP handle provides the best balance of properties *(Table 1.4)*, and yet quality has to be traded off against material and processing costs.

Table **1.4**

Properties of different types of hammer handle

property	wood	steel	GRP
mass	light	heavy	light
damping	good	poor	very good
handle shrinkage	progressive	nil	some initial shrinkage

The relevant British standard (BS 876) is very specific about the type of reinforcement and how the handle shall be manufactured:

'Glass fibre reinforced plastics handles shall be moulded in thermosetting type resins containing a minimum of a 60% glass fibre by volume. Glass fibres shall be continuous, unidirectional and longitudinal... Handles shall be fixed to the head by a chemical adhesive or other means so that that they will not loosen in the head during use.'

Different materials provide different shapes of handle *(Figure 1.3)*, and the ability to mould and self colour direct to final shape has been successfully used in the GRP design.

Conclusion

GRP is an attractive alternative medium to wood or steel, and its technical quality is comparable to (or better than) existing handles.

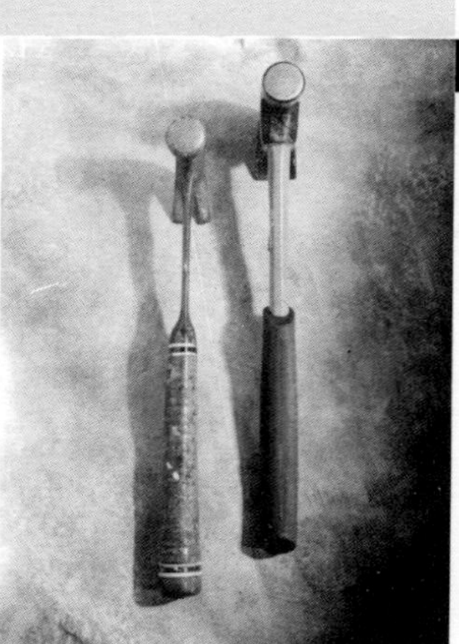

Figure **1.3**

Hammers with different types of handle: in both views steel is shown on the left, glass reinforced plastic on the right

Box 1.3

Manufacture of large lightweight structures

Wind turbine blades have to be large to extract useful power from a low-density medium such as air. Moreover, the blades have to be aerodynamically shaped, smooth, light, dimensionally accurate and capable of sustaining a very large number of load cycles (approaching 500 million) over their projected life of 30 years.

Glass reinforced plastics have been the favoured material from the start of the industry in its present form in the mid-1970s. The main alternative material is timber which is a natural composite and exhibits similar characteristics and properties. To build such large structures cheaply, boat building techniques and materials such as glass fibre reinforcement and polyester resin were adopted.

A modern blade typically comprises only three parts – the outermost shells and a load-bearing spar *(Figure 1.4)* – and each component can be separately optimized and manufactured according to the shape and loading it sustains.

Processing

The outer surfaces must be aerodynamically smooth and this is achieved by laying up the shell sections in a mould of the appropriate shape, the outer surface being in contact with the mould surface. The cross-section of the blade is shown in *Figure 1.4*.

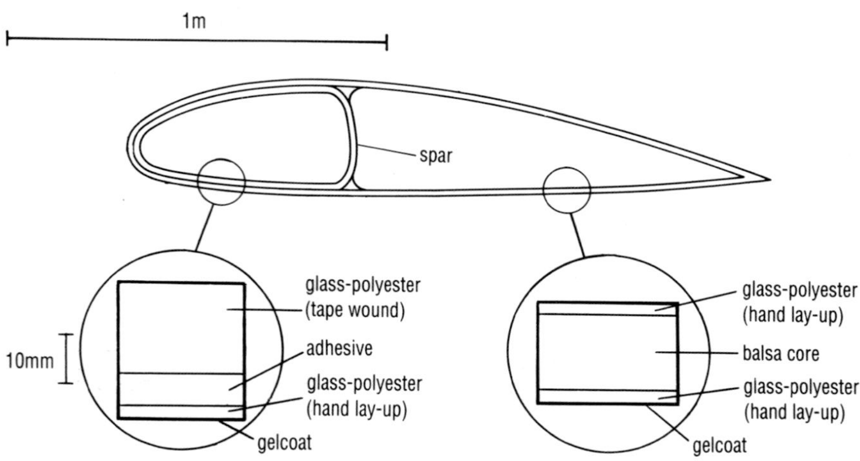

Figure 1.4

Cross section of the blade developed for the wind turbine at Nibe, Denmark, which has been designed for volume production of blades up to 30 m in length (Lilholt)

The mould in which the shell is laid up accurately represents the desired shape of the blade *(Figure 1.5)*, and fabric reinforcement can be selected so that the majority of the fibres (about 90%) lie along its longitudinal axis.

The spar, on the other hand, can be manufactured using a tape winding technique *(Figure 1.6)* in which the tape is impregnated with resin prior to being wound onto a mandrel. The glass fibres are orientated transverse to the length of the tape so that they lie at only a small angle to the length of the finished spar.

Conclusion

Glass reinforced plastics are ideal for this type of application as it is possible to manufacture large blades rapidly and easily. This has enabled the wind turbine industry to develop rapidly over a short period of 10 years. There have been no problems to date with moisture, corrosion or erosion from rain, sea spray or sand particles as might have occurred with other materials.

The potential disadvantages summarised in *Table 1.3* are now briefly discussed.

Figure **1.5**

Shape of blade shell showing taper and twist (Lilholt)

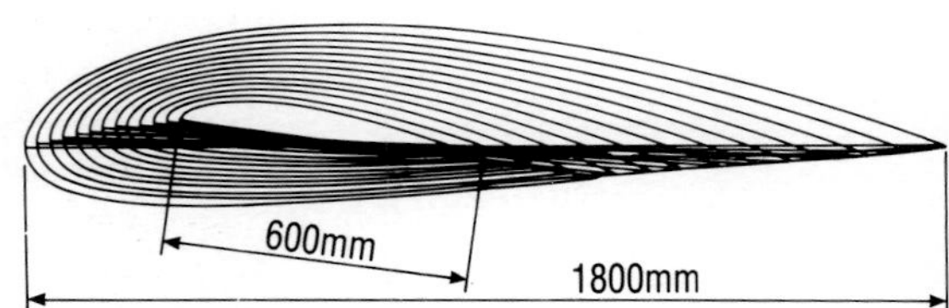

Properties are not derived until the component has been fabricated and they are dependent upon the correct selection of materials, fabrication and cure. The designer should seek advice from materials suppliers and moulders before using an unfamiliar combination of materials.

The **limited availability of design data** stems in part from the large number of combinations of materials and processes that are possible. Material options are set out in *Table 1.5*, and these vary from working with stock items through to using raw materials which, once characterized, should pose no further problems. The best documentation is available for short fibre injection moulded thermoplastics and continuous fibre prepreg systems *(Box 2.3)*.

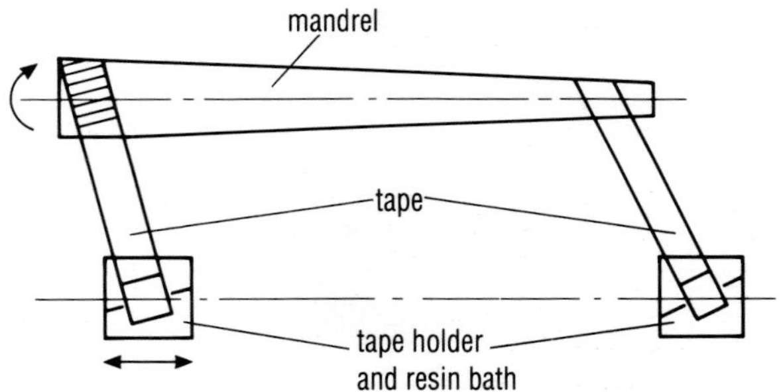

Figure **1.6**

Winding a tape onto a mandrel to form the D-section spar showing tape holder and resin bath at extremes of travel (Lilholt)

Table **1.5**

Material selection options

item	material combinations	shape/size range	reference (section)	type
stock items	fixed	fixed	2.4	mouldings
intermediate materials and compounds	fixed	variable, some restrictions	5.4	prepregs, moulding compounds
raw material	variable	variable	5.1, 5.3	fibres, resins

The **incorrect location of reinforcement** can lead to local weaknesses as the contribution of the remaining fibres to the plastic resin (or matrix) may not be adequate. This can result in premature failure even if the overall level of reinforcement is satisfactory. Options for improving fibre distribution range from using continuous rather than short fibres to using stock items in which the fibre distribution has been specified by the manufacturer. Reference is given in *Table 1.6* to where these are discussed in the text.

Table **1.6**

Options for improving fibre distribution

options	**reference** (section)
use of stock items	2.4
intermediate materials and compounds	5.4
suitable design of shape and form	3.1
suitable design of mould	6.2
use of preform fabrics	6.4
use of continuous fibres	5.2

Standards and codes are not available for many applications, due to the evolving nature of the technology. However, there are a number of general design standards *(Section 4.6)*, and others have been developed for specific applications *(Section 4.7; Box 1.2*).

Recycling reinforced plastic products is difficult and it has only recently been achieved for specific materials *(Section 3.9)*. The onus is therefore on the designer to use the material as efficiently as possible and design for a long life or design for recycling. This is discussed from various aspects throughout this text as both efficiency and long life are possible using these materials.

Public acceptance of reinforced plastics is generally favourable, partly because of the way in which the raw materials have enabled innovative designs to be established, and partly because engineers have tended to be cautious in using them for load-bearing components.

One aspect that could alter this favourable perception is the increasing amount of plastic waste (generated mainly by the packaging industry), which is only just starting to be recycled in appreciable quantity. Recycling of all plastics, including reinforced, is becoming a major concern and numerous initiatives are under way, particularly in Germany *(Section 3.9)*.

Fire, toxicity and smoke hazards can be a problem as some plastics burn intensely and give off both smoke and toxic fumes. However, other plastics such as phenolics and certain thermoplastics have low emissions of smoke and toxic gases, and fire retardancy can be enhanced in other resins by the use of suitable additives *(Section 4.9)*.

1.4 Design brief and conceptual design

Product design begins with a design brief and a performance specification. The way in which these are compiled by the client, preferably in consultation with the designer, will often determine how innovative the product can be. The simpler the brief, the more scope there will be to develop the design.

The initial step in any design *(Section 2.1)* is to establish the design options by ascertaining the relevance of any essential requirements, codes or standards for the product *(Sections 4.2 to 4.4)*. From this set of options, a conceptual design (or designs) can be derived by considering the design requirements of each component, (see *Table 1.7*).

1.4 Design brief and conceptual design

Table **1.7**

Principal design considerations of a product

principal design considerations

- new or improved product
- functional requirements
- shape, colours and texture
- operating environment
- performance
- product life
- material type
- production volume
- product cost
- environmental impact

Considering each in turn:

Product innovation or improvement. Incremental design changes are easier to make with reinforced plastics than with metals because colours and reinforcement can be changed. However, to take full advantage of the properties of such materials, the initial design needs to be undertaken in reinforced plastics.

Requirements can be separated in the design brief into three groups – namely functional, ergonomic and aesthetic – and these need to be considered for each component. Many of these aspects of reinforced plastics have already been outlined, including material properties, ability to integrate parts, moulding to final dimensions, the efficient use of materials with little wastage, and the ability to produce lightweight components with high strength and stiffness.

Shape, colour and texture can all be achieved readily with reinforced plastic materials, and they are often used to give unique design features to products.

Operating environment will involve factors such as corrosion, erosion and humidity *(Section 3.7)*. Though reinforced plastics have attractive properties, past experience and advice from the manufacturer has to be used as a guide in deciding to use these materials. High-temperature operation is a particular concern because the choice of resin becomes restricted as the temperature increases *(Section 5.3)*.

Performance concerns the loading that a component will encounter under all normal (and abnormal) service conditions, and thus an understanding of the response of material types to loads is important. For reinforced plastics, this response can be designed into the material to a large extent *(Sections 3.4, 3.5)*.

Consideration may also need to be given to working practices. For example, clearance holes must be fully aligned before a bolt is inserted as reinforced plastics do not deform plastically. Further examples are described in later sections.

Long **product life** is possible with reinforced plastics for the following reasons:

- incremental changes in design can be readily incorporated as a result of the ease with which simple changes in styling, texture or colouring can be made at low cost.
- mechanical properties such as stiffness can be varied by altering either the type of fibre or the fibre volume fraction without necessarily altering tooling.
- corrosion resistance is unlikely to be a problem if appropriate material selection, manufacturing procedure and quality control have been used. The excellent condition of an underground gasoline tank after many years of service is described in *Box 1.4*.

Material type selection is usually governed by the preceding design considerations, as some materials will fulfil these better than others. If a choice is still possible, then it is often best to offer the client two (or more) conceptual designs.

Production volume. Manufacturing processes can be selected with reinforced plastic materials so that the manufacturing cost matches the volume required *(Section 2.6)*. It is therefore possible to begin with simple tooling and upgrade the tooling as the production requirement builds up.

Product cost. Though GRP materials are not in general as cheap as metals, the ability to manufacture direct to final shape, colour and texture will reduce manufacturing and assembly costs which, together with the raw material, form the initial cost of the product. On this basis, many products are cheaper when manufactured in reinforced plastics.

Environmental impact of the manufacturing process and the product should be minimised (eco-labelling, *Section 2.10*, and *3.9*). Attention therefore needs to be focused on how the product could ultimately be reused, refurbished or recycled.

Box 1.4

Durability of a gasoline storage tank

An underground gasoline storage tank was unearthed in 1989 at the site of a former service station in Chicago, where it had been in continuous use for 25 years. According to Ben Bogner of Amoco Chemicals Company, the tank was in excellent condition. 'It still had gasoline in it, but there were no signs of leakage into the soil, structural distress, or corrosive and chemical attack', he said.

Figure 1.7

Composite gasoline tank being removed from the ground after 25 years of service (Amoco/IRPI)

'The tank was strong enough to be lifted from the hole by a crane using the centre outlet pipe as the only source of support (*Figure 1.7*). It is obvious that this tank could have been used for many more years had the site remained in use as a service station.'

The tank was one of about 60 developed in the early 1960s by Amoco Chemicals in conjunction with Amoco Oil. At that time the company was looking for an alternative to steel tanks, which were prone to leaking due to corrosion. Nowadays corrosion resistance for underground gasoline and chemical storage tanks is mandated by the US Environmental Protection Agency whose rules hold companies burying storage tanks financially responsible for cleaning up any leakage.

The exterior of the 25-year-old tank was in excellent condition. No leaks were noted even though the soil was saturated with moisture to about a third of the way up the tank wall. There was no internal corrosion in spite of the fact that the fuel on the bottom contained moisture.

Conclusion
'We unearthed two steel tanks from the same site and both were severely corroded,' added Bob Bogner. 'While several techniques are available for retarding steel tank corrosion, this 25-year-old composite tank demonstrates the obvious benefit of the inherent corrosion resistance of iso-polyester tanks,' he concluded.

If at any stage it is not possible to fulfil the requirements of the design brief or performance specification, then the preceding design considerations must be reiterated or the requirements re-evaluated with the client (*Figure 2.2*).

An example of how a design brief was successfully converted into a conceptual design is described in *Box 1.5*.

Market niches

Materials tend to be used where either the economics are favourable or else there is an advantage to be obtained in terms of aesthetics, ergonomics or function.

Some of the major niches at present for reinforced plastics include:

- household appliances such as jug kettles and irons, for functional, aesthetic and ergonomic reasons (design).
- sports goods such as fishing rods, skis and tennis rackets, for function (stiffness).
- storing and transporting corrosive liquids, for function (corrosion resistance).
- motor vehicles, for aesthetics (shape, aerodynamics) and function (integration of parts, reduction in mass).
- blades for wind turbines, for both aesthetics (shape, aerodynamics) and performance (reduction in mass).
- rail transport, for function (reduction in mass).
- aerospace, for function (integration of parts, reduction in mass).
- boats, for aesthetics and function (resistance to sea water).

Box 1.5

Conceptual design of an inertial energy storage unit

The **brief** was to reduce the fuel consumption of a city bus by at least 25% by storing the kinetic energy that would otherwise be wasted on braking the vehicle and on stopping to set down or pick up passengers. The stored energy should then be able to be reused to accelerate the vehicle, thereby reducing fuel consumption and pollution. Safety and reliability were of prime importance and the performance specification is set out in *Table 1.8*.

Table 1.8

Performance specification of an inertial store

aspect	requirement
safety, reliability	high
capacity	0.4 kWh
power	160 kW
power loss	< 3 kW
volume	Figure 1.8
mass target	100 kg
stop/start cycles	> 500,000
payback time	3 years

A device storing energy by virtue of its inertia, that is a **flywheel**, seemed the most suitable in terms of its power to mass ratio. Both the mass limit and large number of duty cycles necessitated a material which was both light and strong. The ratio of strength to density *(Table 2.10)* is an important consideration and GRP is four times better in this respect than steel *(Table 2.7)*. Moreover, GRP offered a benign failure mode *(Section 3.4)*, so it was selected.

The restriction on volume *(Figure 1.8)* meant that the design had to proceed from the outside inwards. Lightweight materials permitted the use of a double casing in which the outer could be optimized for stiffness and toughness and the the inner for energy absorption.

The **energy** stored in a flywheel depends linearly upon its mass, but upon the square power of its radius and angular speed. Since the radius in this case was fixed by the space envelope and a flywheel's mass is most effective if concentrated near its periphery, a cylindrical shape was chosen *(Figure 1.9)*. With the dimensions fixed, the

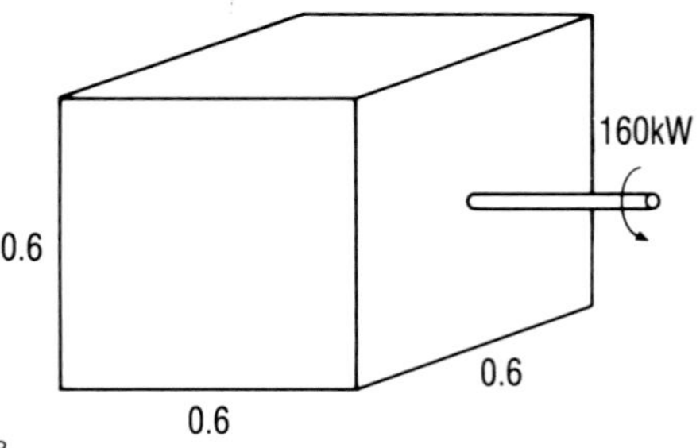

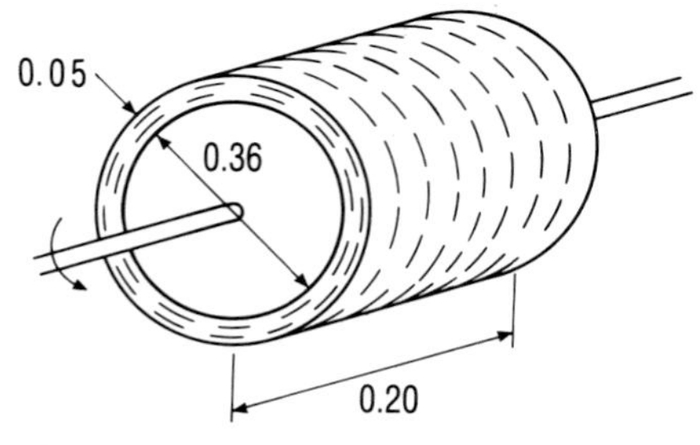

Figure 1.8

(a) Space envelope (m) and power output of flywheel assembly
(b) Rim dimensions (m) of flywheel rotor

maximum rotational speed was set by the stored energy requirement, and the fatigue strength of the material *(Section 3.5)*.

The conceptual design specification is set out in *Table 1.9*.

Conclusion
The device was technically possible and reinforced plastics the only materials to meet the brief. The design was subsequently refined and prototypes successfully tested.

Table **1.9**

Outline specification of a flywheel

parameter	magnitude
outer diameter	600 mm
overall length	560 mm
rim outer diameter	460 mm
rim inner diameter	360 mm
rim length	200 mm
rim mass	25 kg
maximum rotational speed	16,000 rpm
mass available for rotor support, shaft and casings	75 kg

1.5 Conceptual design of a housing

To show how this guide can be used, consider the example of designing a housing to contain an electric machine, starting with the design brief and proceeding through the various steps to the development of a conceptual design.

Design brief
Suppose that the client would like to develop a new type of housing to contain an electrical machine. Market research indicates that the market could grow to 100 units a year within three years and engineering staff would need to develop a few preproduction models and test these under service conditions with interested customers.

The design requirements are set out in Table 1.10.

Table **1.10**

Design requirements of a housing

requirements

- internal size 0.8 m on each side
- external use in exposed location
- mount on wall or pole
- support a heavy internal load
- long life
- access for maintenance
- vandal resistant
- electrical connections
- production target 100 a year within 3 years
- low cost

Design options

To convert the brief into design options, the initial step *(Figure 2.1)* is to consider the statutory directives and whether any standards or codes are applicable *(Section 4.3)*. Safety and liability issues then need to be identified *(Sections 4.1, 4.2)* and the design options addressed. These might include:

- including an inspection hatch for ease of access and to facilitate maintenance and checking of the installation, both mechanical and electrical.
- including an opening at the top of the housing to facilitate heavy maintenance or even replacement of the equipment.
- making the material of the housing electrically insulating so that an electrical short circuit cannot make the housing live, or (less preferred) fitting an insulating liner inside the housing.
- ensuring that the machine cannot fall through the bottom of the housing as a result of environmental corrosion.
- preventing the housing from falling off its support.
- using materials that do not corrode to ensure that the housing, brackets and method of support have adequate strength to support the load.

The relevant location of information is summarized in *Table 1.11*.

Conceptual design

To establish a conceptual design *(Figure 2.2)*, the next step is to consider the optimum number of components and to identify the function that each has to perform *(Table 1.12)*. In making such an analysis the designer must remember to locate joint lines in positions where they will be lightly loaded *(Section 2.7)*. One such concept is shown in *Figure 1.9*.

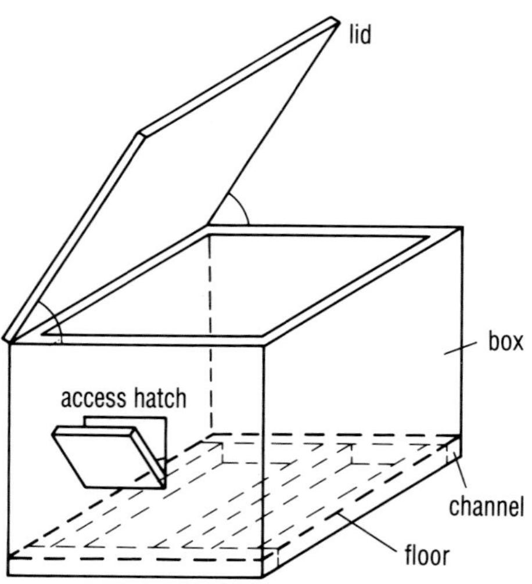

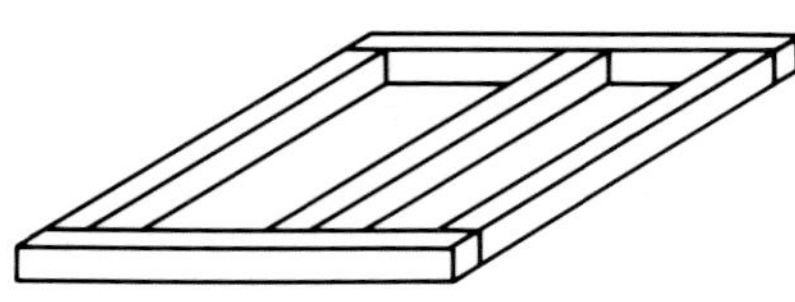

Figure 1.9

Conceptual design of container showing location and proposed mounting of lid and access hatch. One option would be to make the floor from stock items using box sections as the load-bearing member to which a flat sheet is secured

Material choice

The material requirements are summarized in *Table 1.13*, together with the prime reason for each. All these requirements could be satisfied by reinforced plastics.

Table **1.11**

Safety and liability aspects

design aspect	reference (section)
directives and standards	4.4
quality systems	4.5
identify safety issues	
maintenance	
access and replacement	
electrical short circuits	4.3
identify liability	
load falls out of housing	
housing falls off support	4.2

Table **1.12**

Component functions

	functional	ergonomic	aesthetic
lid	vandal-proof insertion and removal of machine	light, easy to open	
hatch	access to electrical connections	serve as tray for tools	
box	vandal-proof, electrical insulating	easy to maintain	neutral colour
floor	strong	lifting eyes for ease of mounting	

Table **1.13**

Material requirements

design requirement	reason
light	easy to lift
strong	to sustain load
tough	vandal-proof
self-coloured and non-corrosive	low maintenance, long life
electrical insulating	safety

Manufacturing route

The requirement for a small but growing market could initially be satisfied by using a contact moulding process *(Section 6.2)* and then converting to a closed mould technique once the market had been established *(Section 6.4) (Figure 2.12)*.

However, three of the components could be manufactured from stock items *(Box 2.1, Section 2.11)*, and so only one part, namely the box, would need to be manufactured. For prototype development this too could be made from stock items.

An outline cost could therefore be derived for the conceptual design by discussing the moulding with a manufacturer and obtaining the costs of the stock items from suppliers' catalogues. Advice on jointing could be obtained from the manufacturers of the stock items.

Conclusion

The design is feasible in reinforced plastics and this is probably the best option. The preferred material combination would be glass reinforced polyester on grounds of cost and widest availability of stock items *(Section 2.11)*.

1.6 **Reference information**

Journals

Engineering, Design Council, London – monthly with a bi-monthly supplement on Advanced Composites Engineering.

Eureka, Infopress, Horton Kirby – monthly discussing the use of materials and design in an innovative manner.

PRW, EMAP, Croydon – weekly newspaper covering the plastics and rubber industry.

IRPI, Channel Publications, High Wycombe – bi-monthly trade journal covering the latest developments in the reinforced plastics industry.

Fibreworld, Vetrotex, Chambery and *Fibreglass Facts,* OCF, Bruxelles – trade journals with useful examples of where composites have been successfully used.

Composites and *Composites Manufacturing,* Butterworth, London – technical journals with useful features on aspects such as abstracts of patents and literature survey.

Reinforced Plastics, Elsevier, Oxford – emphasis on recent developments in the technology.

Use in various sectors

Composites in the Automotive Industry, Seminar S845, IMechE, London, 1990.

Composites in the Aerospace Industry, Seminar S844 C S Smith, IMechE, London, 1990.

Design of Marine Structures in Composite Materials, Elsevier, Barking, 1990.

Composites in the Chemical and Processing Industry, Seminar S721, IMechE, London, 1989.

'French market for composites in the automotive industry: situation and evolution', G Buisson, *Fibreworld* No 1 p16, Vetrotex, Chambery, 1989.

Glass Fibre Reinforced Plastics and Buildings, A J Legatt, Butterworth, London, 1984.

Use in industrial products

'Wingblades of glass fibre reinforced polyester for a 630 kW windturbine', H Lilholt et al, *Proc ICCM-3 Conf,* Elsevier, Amsterdam, 1980.

'Energy storage devices based on flywheels', I Duff Barclay and R M Mayer, *Proc Autotech 85 Symposium, Birmingham,* IMechE, London, 1985.

'Excellent condition after 25 years of a gasoline storage tank', *IRPI* Channel Publications, High Wycombe, June1989.

'Success stories', *Advanced Composites Engineering,* Design Council, London, June 1989.

Getting started

Design Data for Fibreglass Composites, OCF, Ascot, 1985.

Polyester Handbook, Scott Bader, Woolaston, 1986.

Composites: a Design Guide, T Richardson, Industrial Press, New York, 1987.

Successful Composite Techniques, 2nd ed, K Noakes, Osprey, London, 1992.

Design Data for Reinforced Plastics, N Hancox and R M Mayer, Chapman & Hall, London, 1993.

Other publications from:
Institute of Materials, 1 Carlton Terrace, London SW1Y 5DB +44 071 839 4071
British Plastics Federation, 5-6 Bath Place, Rivington Street, London EC2A 3JE +44 071 457 5000
Material manufacturers.

Assistance

The Plastics and Rubber Advisory Service will assist with any query or problem related to plastics, including reinforced plastics. It will deal with queries by phone or fax and has its own database of proprietary items, custom and trade moulders and material suppliers. It can also provide advice on production techniques and material selection. 5-6 Bath Place, Rivington Street, London EC2A 3JE +44 0839 506 070.

Material sources

UK Materials Information Sources 2nd ed, K W Reynard, Design Council, London, 1992 – lists sources alphabetically by material type with subject and name indexes.

Materials Data Sources P T Houlcroft, Mechanical Engineering Publications, London, 1987 – reference information includes journals, directories, helpful organizations and which manufacturers supply what products.

Embodiment design considerations

chapter 2

summary

This chapter sets out a design method which takes account of the properties and peculiarities of fibre reinforced plastics. It suggests how to approach the design process, and how to select appropriate materials and manufacturing methods. It then goes on to discuss jointing and costs.

Introduction

The introductory chapter discussed the concept of working from a design brief and performance specification towards a conceptual design. In this chapter a design route for reinforced plastics products is outlined which closely follows that for other materials, as advocated by British Standard BS 7000, *Guide to managing product design*.

From the conceptual design, the various steps in deriving an embodiment design are considered. They should be undertaken in a systematic manner, the time spent on each step depending on the nature of the product and its intended use.

It may also be desirable to consider an alternative design in another material, possibly a metal, if the advantages of reinforced plastics are not obvious.

The embodiment design must be developed far enough for the client and potential users to be able to evaluate the product and its performance. It is on this evaluation that a decision must be based on whether to commit resources and proceed with detailed design and manufacture of prototypes and volume production.

Many of the aspects introduced in this chapter are covered further elsewhere, so readers are advised to consult the relevant sections if more details are required.

2.1 Conceptual design

As discussed in *Section 1.4*, the design brief and performance specification should set out a description of what is required and should contain information on performance, cost and time-scale requirements.

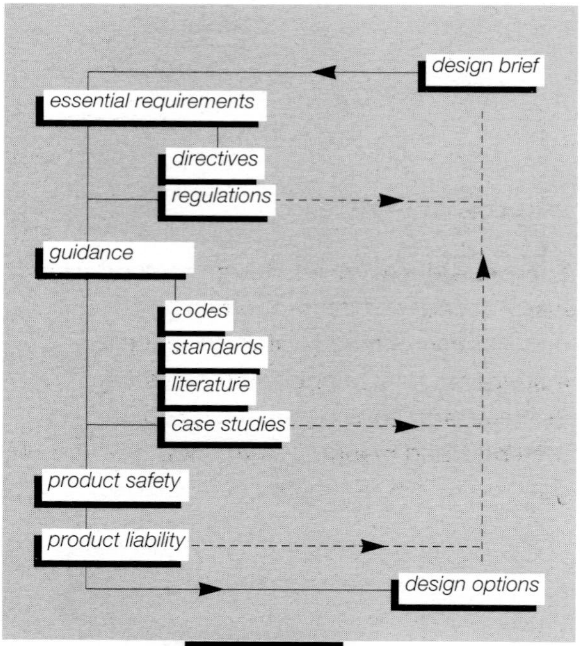

Design options

The design options are established from the design brief by considering both the essential requirements and the guidance available in directives, codes and standards *(Figure 2.1)*. These are considered in some depth in *Chapter 4*. This screening process may uncover reasons for rejecting certain designs, materials or processes.

Figure 2.1

Establishing design options. Solid lines indicate 'go' options and dotted lines 'stop and think' options

Establishing the design

The conceptual design results from a consideration of the design options and the number of components that constitute the product *(Figure 2.2)*.

The choice of components will depend upon materials, production processes and costs, the availability of stock items (*Section 2.4*) and the ease (or otherwise) of making joints. If any of the aesthetic, ergonomic and functional requirements cannot be met, then the number and type of components (or the design options) must be reconsidered, for example by moving a joint or integrating components.

If there are any conflicts at this stage of the design then these must be reconciled with the client before proceeding further.

2.2 Embodiment design

The embodiment (or layout) design is arrived at by considering materials and processes as discussed in *Chapters 5* and *6* respectively *(Figure 2.3)*.

As the design is developed in this sequence, it must be checked for its fitness of purpose. If this cannot be achieved then the conceptual design, materials or production processes involved must be reassessed.

The nature of fibre reinforced materials is such that the selection of processes, materials and properties are closely interrelated, and the choice of any one may define the other two. Though this may restrict the designer's choice, it is beneficial in the sense that it tends to reduce the potentially very large number of possible combinations to a manageable number of options.

The design loop *(Figure 2.4)* is always iterated a number of times – at least once for each of the major stages of conceptual, embodiment and detailed design – and in successively greater depth in order to ensure that there is no fundamental problem which would prevent the design being realized.

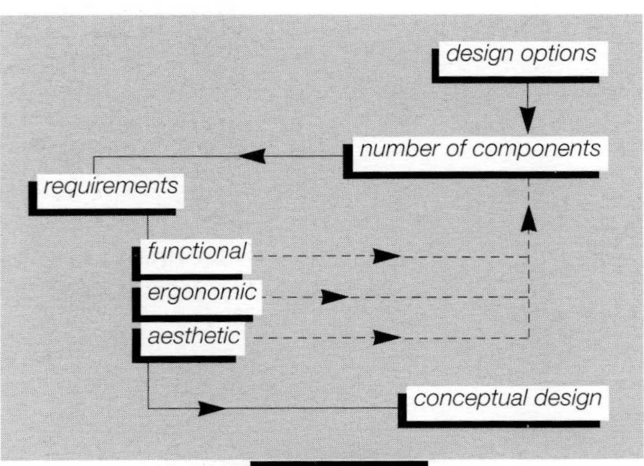

Figure **2.2**

Conceptual design

Validation

The embodiment design should be validated by making a model, mock-up or prototype, depending on which aspects of the design need to be checked. This will ensure that the requirements of the design brief can be met and will allow the client to gauge customer acceptability. Tests with a prototype will allow the designer to ascertain that the materials specified in the design can be employed in practice.

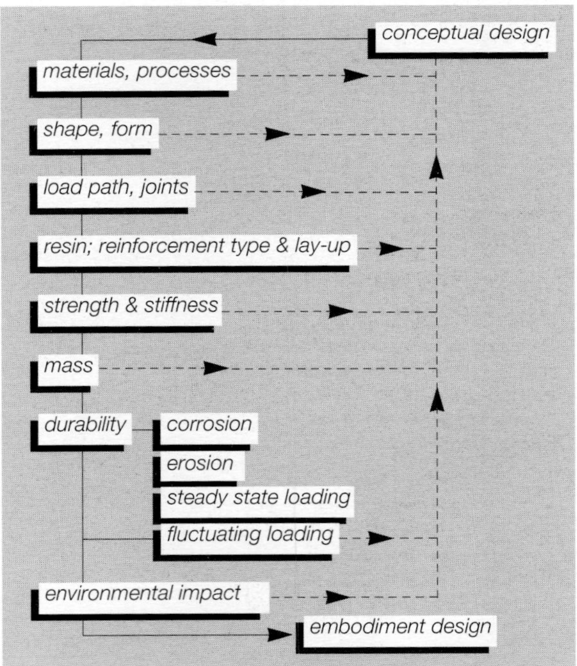

Figure **2.3**

Establishing an embodiment design; this loop must be repeated for the detailed design phase

2.3 Detailed design

The detailed design stage involves, first, incorporating alterations required as the result of validation or any other changes in the design brief. The embodiment design sequence is then repeated in more depth and this forms the basis of *Chapter 3*, together with further guidance on good working practices.

While the hazards of fire *(Section 4.8)* must be considered at each stage of the design, materials selection and aspects of the design must be checked during this stage to ensure that the margin of safety is adequate *(Section 4.9)*; a suitable strategy is set out in *Figure 2.5*. The extent to which fire risks should be investigated depends upon the nature of the application, but they should always be considered.

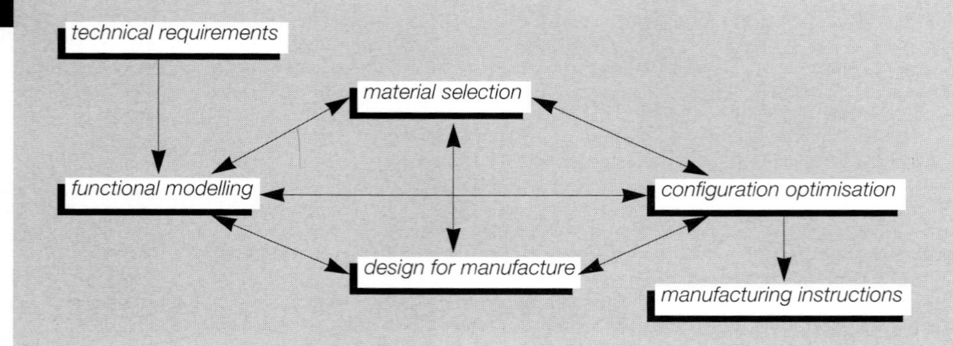

Figure **2.4**

Design loop indicating key steps (Wallace)

Manufacturing route

The production process route is decided at an earlier stage in the design, but quality assurance must be put in place so that the resulting component(s) will conform with the detailed design in all respects. Reference should always be made to agreed international standards *(Box 4.3)*, but these may set only minimum requirements. The designer therefore needs to work closely with materials suppliers and the manufacturer so that the level of quality control is sufficient to ensure both the integrity and performance of the product. Test methods, including those for non-destructive evaluation, are set out in *Chapter 7*.

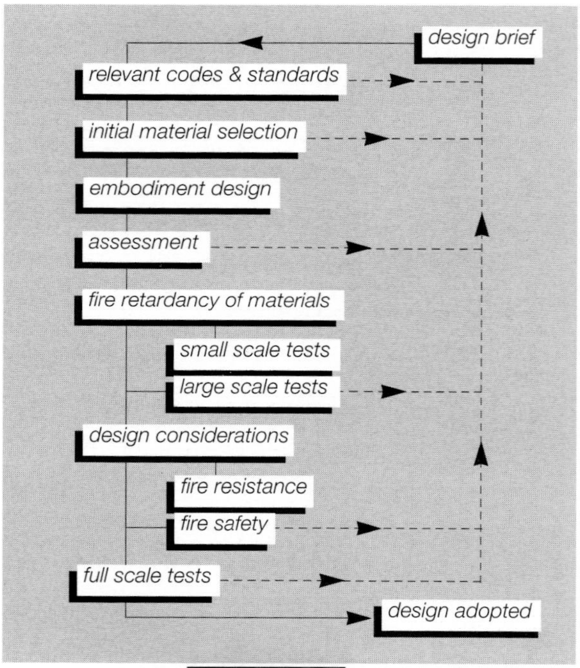

Figure **2.5**

Design strategy for fire prevention; initial screening has to be followed, where necessary, by a detailed assessment

Checklist

BS 7000 provides a formal guide to managing product design, complete with checklists of what management, product managers and designers should consider at each stage of the design process. Information in this guide follows the same stages, with necessary considerations summarized in tables and good working practices described, based on the authors' experience. As designers become more experienced, they will add their own checks.

Simultaneous engineering

The aim of simultaneous engineering is to reduce the lead time between starting a design and manufacturing a product by ensuring that design for production starts as soon after the start of the product design as possible. This is essential for reinforced plastics because of the way in which material properties are only fully realized on manufacture.

2.4 **Materials options**

There are three broad options for selecting materials:

- finished products
- compounds and intermediate materials
- raw materials

Finished products

These are limited to specific shapes, sizes and materials, but their properties are known and are highly reliable. The advantages of using such items are set out in *Box 2.1* and, wherever they can be used, they are the quickest and easiest method of incorporating composite components into a design.

Box **2.1**

Stock items

Availability

Certain sections are available in bar, rod or panel form. They are manufactured by pultrusion, moulding or filament winding *(Box 2.3)* and so are already in their final form. The considerable advantages of using such items are listed in *Table 2.1* and typical range of available products is set out in *Section 2.11*.

Maintenance is low because components can be self coloured and resins are available to withstand highly corrosive environments. Hollow or foamed core sections have high strength and longitudinal stiffness. Such sections are not necessarily isotropic because of the nature of their reinforcement, so the more carefully one can describe what is required the more likely it is that the supplier will be able to meet the requirements.

Table **2.1**

Advantages of stock items

advantages
- known and consistent property values
- high level of mechanical properties
- low scatter
- good surface finish
- wide range of colours
- wide range of shapes and sizes
- long lengths available
- low maintenance

Jointing of stock items needs careful consideration *(Sections 2.7* and *3.8)* as few proprietary systems are available.

Use in structures

A large variety of structures incorporate pultruded sections, including ladders, grids, racking, handrails, soil-stabilising slats, guttering, channels and bridge maintenance platforms.

Flat sheeting is tough, durable and can be textured. It is used in lining walls and ceilings where a hygienic environment is required, in vehicle bodies for strength, light weight and toughness, and for both lightweight transparent and translucent roof sheeting.

It is possible to be quite creative with such components, as the example of a box with a snap-on lid in *Figure 2.6* shows. Structures can also be assembled by using suitable combinations of items such as pultrusions and flat sheets *(Figure 2.7)*.

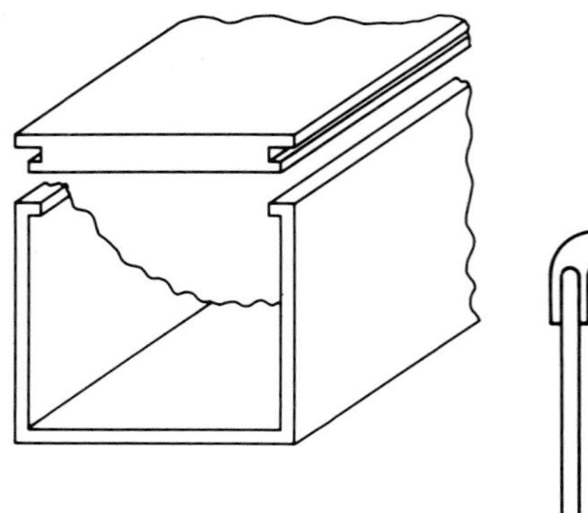

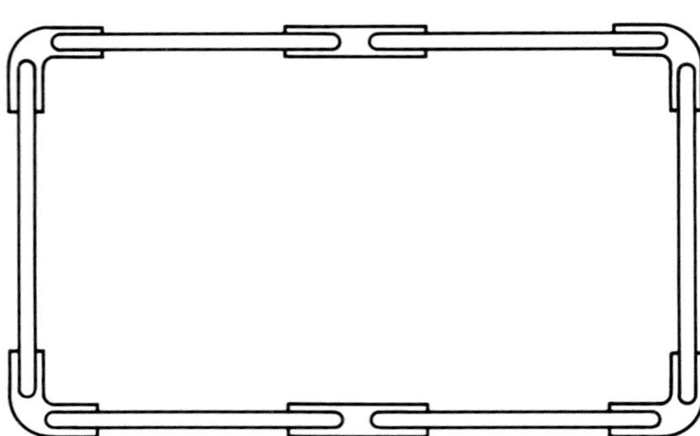

Figures 2.6 & 2.7

Box with snap-on lid. There is sufficient flexibility in the sides of the box to allow the lid to be fitted

Three separate types of pultrusion used to form a rectangular channel section (Fibreline)

Compounds and intermediate materials

With such materials, the combination of reinforcement, resin and additives is preselected by the supplier or compounder and the production process route is prescribed to produce the optimum properties. A limited number of material combinations are available and these are described in *Box 5.8*.

Moulding compounds and glass mat thermoplastics are generally used for high-volume production runs such as those for automotive components, while prepregs and thermoplastic sheet compounds are used where it is important to optimize fibre lay-up and alignment, for example in the aerospace industry (*Box 5.8*).

Raw materials

These provide the largest range of options and materials can be tailored to a wide range of properties and production processes. The major disadvantage is that property levels must be checked to ensure that the reinforcement, resin and additives chosen are all compatible and that the correct process parameters have been used.

2.5 Material properties

The combination of reinforcement and resin needs to be selected in order to provide a range of properties to fit the design requirements.

Some properties are determined primarily by the resin and others by the reinforcement (*Box 2.2*). While the reinforcement and resin can be selected independently to a large extent, the choice may be constrained by the use of a particular fabrication process, as discussed in the next section.

Property attainment

A particular combination of reinforcement and resin will give rise to a particular range of properties and typical numerical data are given in *Section 5.5*. What follows is a brief description of how these properties can be attained. It should be remembered that one can match only one resin-dependent and one fibre-dependent property at any one time, so this can be a tough design choice.

Box 2.2

Properties determined by fibres and resins

Material selection

One method for selecting the most appropriate type of resin and fibre is to identify the critical design requirement for each, and choose the type that will give the highest property level. The level for each of the other properties then needs to be checked to ensure that these will be satisfactory. If not, a further choice of materials will have to be made (*Chapter 5*).

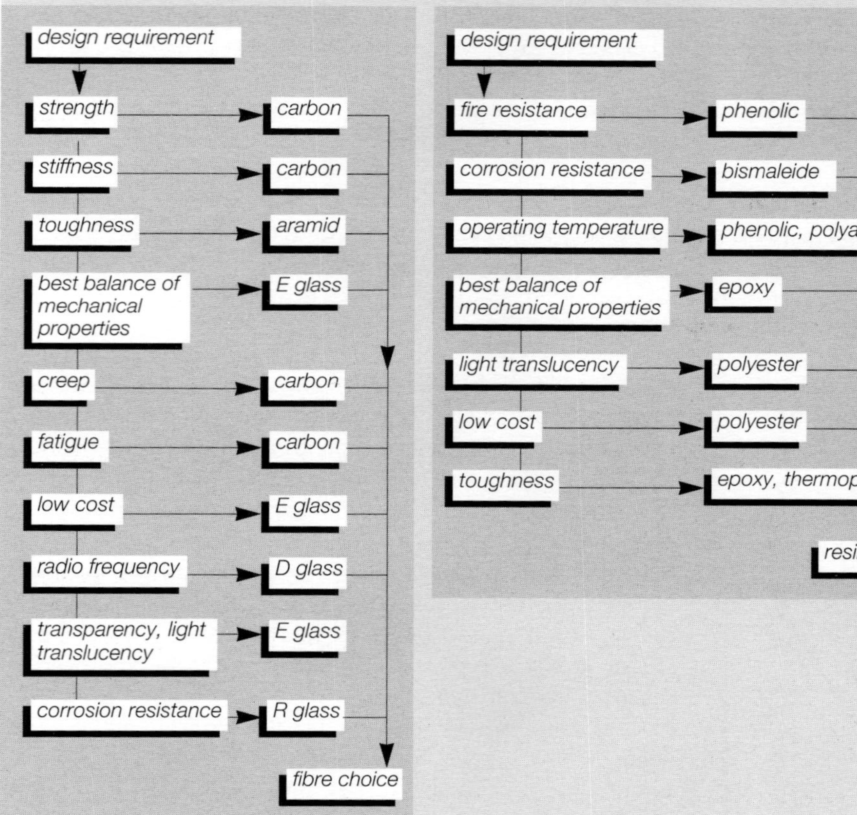

Figures 2.8 & 2.9

Selection of a possible reinforcement and resin

Resin-controlled properties

Fire retardancy. Various levels of properties can be attained, including low smoke and toxic fume emissions. These can be achieved in various ways, from using a resin which is inherently fire retardant (such as phenolic and some thermoplastic resins), to incorporating fire-retardant additives within the resin or coating the component with a fire-retardant paint *(Section 4.9)*.

Corrosion resistance is almost entirely dependent upon on the resin used because it protects the fibres. The fibres themselves will have different corrosion resistance *(Table 5.2)*. In general, thermosetting resins have good resistance to all the major chemical groups with the exception of strong oxidants *(Section 3.7)*.

Temperature resistance depends upon the softening point of the resin. Many resins are available which can withstand up to 125°C, a few can go to 250°C, with an upper limit of 300°C at present *(Section 5.3)*. The properties of glass fibre reinforced resins are not degraded and in fact are usually improved at low temperatures *(Section 7.7)*.

Hardness and **abrasion resistance** are determined by those of the resin together with the fibre. The common thermosetting resins such as epoxy, polyester and phenolic are all intrinsically hard, and this can be increased by the use of suitable fillers.

Thermal conductivity is generally poor, but it can improved by the addition of a conductive filler like aluminium flake or by using aluminium-coated glass fibres.

Electrical insulation is an intrinsic property of some composites, which is why FRP is widely used to make such items as printed circuit boards. Care may be needed with CRP as carbon fibres are conductive *(Table 5.1)*.

Fibre-controlled properties

Properties determined by the fibres differ in two fundamental ways from those dominated by the resin. They depend upon both the amount of fibre (that is, its volume fraction) and its orientation. The degree of anisotropy of the fibres depends upon the fibre length. For short fibres the properties are essentially isotropic, while for longer fibres the properties become increasingly anisotropic.

Tensile strength of a glass reinforced plastic versus glass content (by weight) in the fibre direction (Owens Corning Fiberglas)

Figure 2.10

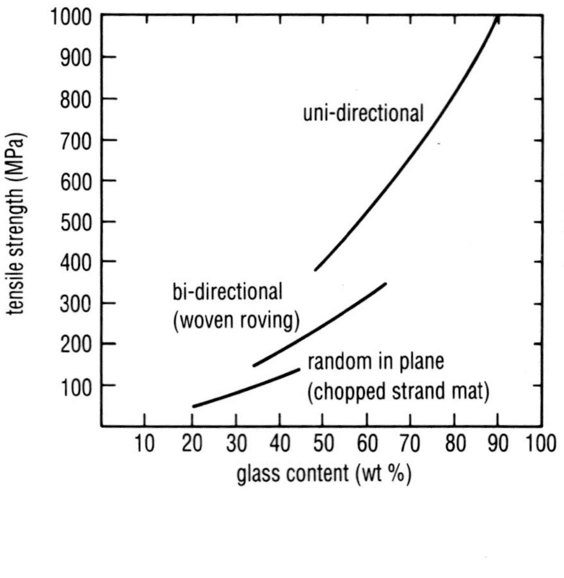

Strength and stiffness variation with fibre volume fraction in the fibre direction is illustrated in *Figure 2.10*. The designer can clearly tailor these properties to the required level, as discussed further in *Box 3.2*. Strength and stiffness can be improved by changing the fibre type and substituting carbon or aramid for glass as these are stronger, stiffer *(Figure 5.1)* and of lower density *(Table 5.1)*.

Toughness is a feature of most fibre reinforced plastics because of the weak interface between the fibre and the resin. Energy is naturally dispersed away from the crack tip, either by fibres debonding or by pulling out or by the crack propagating by delamination between fibres or plies *(Section 5.5)*. However, whereas a metal may deform under a certain impact, reinforced plastics are more likely to craze as energy is dissipated away from the point of impact.

Impact resistance varies with the type of reinforcement (mat, fabric etc) and fibre type so this property can also be tailored. This can be enhanced by adding elastomeric additives to some resins.

Fluctuating loads (fatigue). The response of glass reinforced plastic to cyclic loading is such that a reduction in strength of about 10% per decade can be assumed within the fibre direction *(Section 3.5)*. Aramid and carbon fibres provide a higher level of fatigue resistance *(Box 3.3)*.

Thermal properties. The range of thermal properties is more limited than that of mechanical properties *(Table 5.2)*. They can be matched to some metals though this relies on the level of local reinforcement. Carbon fibre and aramid laminates can have zero or negative thermal expansion coefficients.

Translucency is possible by using a suitable resin and glass fibre sizing and achieving a good wet-out between the two components. Moreover a wide range of heat and light transmissions can be achieved by adding suitable fillers to the resin.

2.6 **Production process selection**

Selecting the right production process is the key to producing the desired number of parts at an acceptable cost, and a variety of processes are available *(Box 2.3)*. Some of these have already been mentioned in the previous chapter.

Box **2.3**

Major fabrication processes

Contact moulding (COM) provides one good surface – that in contact with the mould surface. Large mouldings are possible such as the hulls of boats up to 60 m^2. Tooling costs are low, but the process is generally labour-intensive and skill-dependent.

In the **hand lay-up** method, a laminating resin and individual layers of fabric or mat reinforcement are laid inside a mould and wet-out achieved by rolling the resin into the fibres. This process has been mechanized for manufacturing the hulls of minesweepers and other large vessels. In the **spray up** method, preset proportions of resin and chopped fibre are sprayed into a mould and consolidated using a roller.

Filament winding (FIL) is used primarily for cylindrical components. Reinforcement may be either in the form of tape or roving or a combination of both. Length is no problem up to 30 m (turbine spar, *Figure 1.5*) and diameters up to 6 m have been wound for storage tanks.

Resin transfer moulding (RTM) uses a closed mould to give close tolerances on all surfaces. It has a higher initial cost of tooling than contact moulding, but is much less labour intensive in volume production. The method is highly reproducible and quick-setting resins can be used to reduce manufacturing time. The largest part in volume production, for the Renault Espace car (*Figure 1.2*), is 4.2 m long and about 1 m wide. ▶

Pultrusion (PUL) produces a product of regular cross section by pulling a combination of resin-impregnated rovings, mats and/or fabrics through a die. It is a continuous process which is cheap if suitable tooling already exists. The maximum dimension is about 1 m.

Injection moulding (IM) is a high-volume process which is widely used for manufacturing plastic components. The addition of reinforcing fibres is used to improve the level of properties. The upper size limit is 1 – 2 m^2, though larger parts may be possible using special tooling

Prepreg moulding (PPM) uses pressure or vacuum to consolidate sheets containing fibre preimpregnated with resin *(Section 5.4)*. It is widely used in the aircraft industry and for other highly stressed components.

Press moulding (PRM) of moulding compounds or intermediate materials *(Box 5.8)* can be in either a cold or heated press. The normal size limit is about 3 m^2.

Centrifugal moulding (CEM) involves rotating a mould and spraying chopped fibre/resin mixture onto its inside surface so that centrifugal forces consolidate the mixture. It is suitable for circular-geometry, high-volume components with an upper size limit of 4 m diameter.

For more details, refer to *Chapter 6*.

Most of the manufacturing considerations in *Table 2.2* are self explanatory, but one should note that some of them limit the choice of process, which in turn limits geometrical shape and size. Moreover, with some processes, a resin-rich surface coat (gel coat) can be provided which not only provides a good surface finish but also a ready means of pigmentation. Inserts can also often be incorporated, ranging from fasteners to latches and brackets.

Table **2.2**
Manufacturing considerations

considerations

- size
- shape/form
- number required
- tooling cost
- reinforcement type
- fibre fraction
- gel coat
- inserts

The options for the major processes are summarized in *Table 2.3* and discussed in more detail in *Section 6.1*.

Price and volume

The price/volume relationship is illustrated schematically in Figure 2.11. The price per part decreases as production numbers increase, reflecting the decreasing cost of amortising the tooling and process machinery over a larger number of parts. The figure also shows that specific processes are suitable for specific volume production because of the more elaborate (and expensive) plant required for greater production.

Table 2.3

Typical process options for various manufacturing processes

aspect	COM	FIL	RTM	PUL	IM	PRM	PPM	CEM
size	large	large	medium	small	small	medium	medium	large
shape	any	cylindrical	any	regular	any	any	any	cylindrical
number	small	any	medium	large	large	medium	small	medium
gel coat	yes	yes	yes	no	no	no	yes	yes
inserts	yes	yes	yes	no	no	no	yes	no

Note: for further information refer to Tables 6.1 to 6.3.

COM – contact moulding; FIL – filament winding; RTM – resin transfer moulding; PUL – pultrusion; IM – injection moulding; PRM – press moulding; PPM – prepreg moulding; CEM – centrifugal moulding

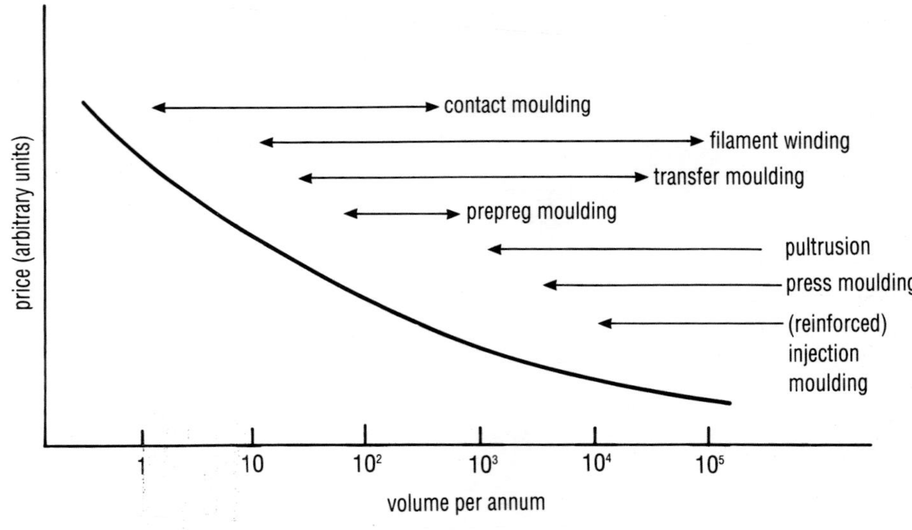

Figure 2.11

Schematic price/volume relationship for various fabrication processes

2.7 Shape, form and joints

Following the selection of processes and materials, the designer can consider further the function and performance of the components derived in the conceptual design stage.

The use of complex shapes, including double curvature and changes in thickness, is possible with these materials either by selecting a fabric with a large amount of compressibility or shape *(Section 5.2)* or by designing the mould in such a way that preferential fibre flow does not occur in certain directions. Use of complex shapes may enable the number of parts to be reduced.

Joints

The standard types of joint can be used provided that adequate attention is given both to the location of the reinforcement and the nature of the surface. Adhesive and bolted joints are principally used, and the respective advantages of each are summarized in *Table 2.4*. Though these considerations apply to all materials, it is the ease with which the joint can be prepared and made, and the possible need for taking it apart, that tends to settle the decision about the type of joint to be used.

Table **2.4**

Comparison of the properties of adhesive versus bolted joints

design aspect	bolts	adhesives
stresses	localized	more uniform
joint stiffness	decreased	increased
sealing	difficult	easy
preparation	drilling holes	prepare surfaces
permanence	removable	fixed
quality control	easy	difficult

Bolted and adhesive joints are discussed in *Boxes 2.4* and *2.5*.

Box **2.4**

Bolted joints

Bolted joints can be as good as in metals provided that there is adequate reinforcement around the bolt hole and the mating surfaces are flat. Good quality moulded surfaces are adequate and it should never be necessary to machine surfaces as this will remove the resin-rich surface layer and expose the underlying fibres.

Holes must not be drilled too close to edges (6 mm minimum) as there may not be enough fibres to prevent bolt pull-out. Extra precautions may be needed with reinforcements other than glass as there are fewer fabrics available which are suited to bolted joints. The prime requirement is to ensure that there are fibres in the principal load directions so that there is adequate strength to resist the compressive and shear forces in the vicinity of the bolt hole. It is necessary to check that creep relaxation does not require the bolt to be retorqued after any initial relaxation. If necessary a larger and/or wavy washer should be considered.

A typical bolted joint

Figure **2.12**

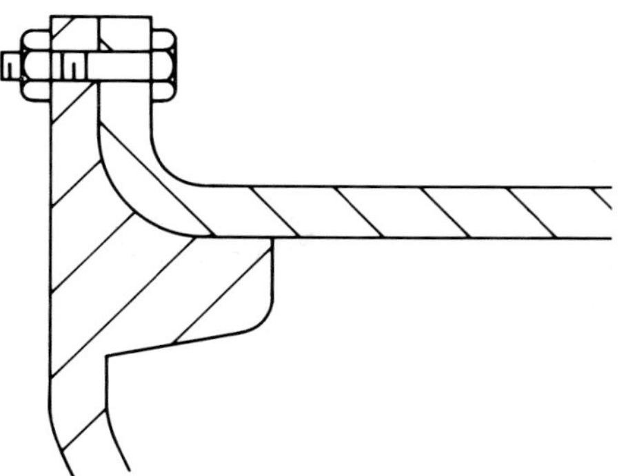

The higher the quality of the joint, the better the level of quality control and testing that will be necessary. Non-destructive evaluation is not as easy as for metallic joints, so adequate inspection is essential *(Section 7.11)*.

A typical joint is shown in *Figure 2.12*.

Location

It is important to locate joints away from highly stressed areas, particularly if adhesive bonding is used, as bond strengths are generally lower than laminate strengths in the through-thickness direction.

Box **2.5**

Adhesive joints

Adhesive joints

A wide range of adhesives are available. Selection depends on specific criteria, the most important of which are set out in *Table 2.5*.

Good design is important and attention should be given to how the loading will occur and the ability of the joint to withstand load. An example of good and bad practice is given in *Figure 2.13*.

The major types of adhesive are listed below, together with some typical applications. Sometimes the resin used as the matrix material can itself be used as an adhesive.

Table **2.5**

Criteria for adhesive selection

selection criteria

- temperature range
- operating environment
- resistance to crack propagation
- whether parts can be clamped during curing
- similar or dissimilar materials to be jointed
- acceptability of adhesive mixing
- heating permitted for curing

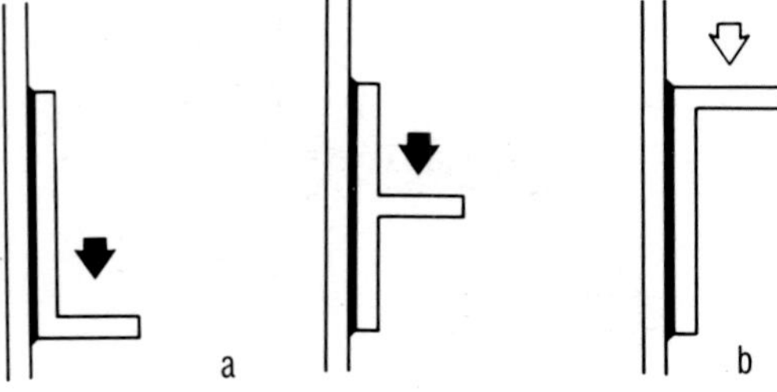

Figure **2.13**

An adhesive joint between two casings (a) good and (b) bad practice (Permabond)

Anaerobics are available in a wide range of viscosities. They cure when treated parts are assembled and air excluded from the mating surfaces and are used for joint sealing and retention of coaxial components.

Cyanoacrylates are low viscosity fluids which cure rapidly when parts are brought together. They are used for assembly of small components generally non-metallic.

Hot melt. Various types of thermoplastic resins are available in a variety of forms which melt on heating and solidify on cooling to form strong bonds. They enable the rapid assembly of lightly stressed components and are good gap fillers.

Epoxies are thermosetting resins. Hot or cold cure are possible and they have good structural properties.

Polyurethanes are usually two components which cure on being brought together. They are generally of high viscosity; strong and resilient, and are widely used in bonding structures. ▶

Phenolics are either thermoplastic or thermosetting resins. The former are cured by heating and cooling; the latter with catalyst. They have good heat resistance and durability in harsh environments.

Toughened adhesives contain dispersed, resilient rubber particles to provide enhanced shock resistance and durability. They are available as anaerobics and epoxies.

Toughened acrylics are based on a variety of acrylic monomers with a range of viscosities. They cure at room temperature, are very versatile and will bond most substrates.

Selection should be based on manufacturers' recommendations. Computer-aided selection is now also possible *(Section 2.10)*.

Screw threads can be cast into reinforced plastics though specialist fabrication techniques are required. Self-tapping screws can be used but are not as strong as in metals. Wire thread or other metal inserts should be used when screws have to be removed and replaced.

Fasteners and inserts are available, some of which are specifically designed for reinforced plastic materials.

Welding can be used for joining parts made from thermoplastic resins. The strength of this joint is that of the matrix alone.

2.8 Embodiment design

With the information provided so far, supplemented by reference to other chapters or sources where necessary, it should be possible to refine the conceptual design by iterating the design loop (*Figure 2.4*) and following the steps set out in *Figure 2.3*.

The initial choice of material and production process needs to be confirmed or altered. This leads on to further consideration of the size and form of each component, their relationship to one another and how they can be attached.

Load path

The next stage is to ensure continuity of the load path. This requires the fibres to be dispersed evenly throughout the volume that is carrying the loads. It is good practice to consult the moulder about the proposed shape of the components so that problems do not arise during manufacture with the flow of short fibres or the movement of longer ones.

Transfer of load across joints and ultimately to 'ground' requires some careful thought as to the type of joint and the similarity of the materials each side of the joint line. Joints should not be located in highly stressed areas.

One can initially assume that one of three levels of reinforcement will be used and that the properties of the resulting composite will reflect the level of reinforcement *(Box 2.6)*. As can be seen from *Table 2.7*, the strength of glass fibre reinforced plastics is closer to that of metals than is their stiffness. As a result, designs with glass fibre reinforced plastics are generally limited by their stiffness rather than their strength.

Box **2.6**

Strength and stiffness

Strength and stiffness

To simplify matters at the embodiment design stage it is possible to consider three levels of reinforcement which correspond to three general types of composite.

Table **2.6**

Typical levels of reinforcement

reinforcement level	**type**	**directionality in plane**
low	short fibres, random mat	omni
medium	coarse fabric	bi
high	fine fabric	bi

These translate into the following properties along the principal fibre directions, using the data from *Section 5.1* and *5.3* and the rule of mixtures *(Box 3.2)*.

Table **2.7**

Typical composite properties for various levels and types of reinforcement compared with an aluminium alloy and steel; fibre volume fraction (V_f), tensile strength (σ), modulus (E) and density (ρ)

material	V_f (%)	σ (MPa)	E (GPa)	ρ (Mg/m^3)
E glass reinforcement:				
low	15	140	8	1.4
medium	30	250	15	1.6
high	50	450	30	1.8
other fibre reinforcement:				
aramid (HM)	30	370	22	1.3
carbon (HS)	30	500	45	1.4
metallic alloys				
aluminium alloy	–	300	70	2.8
steel	–	500	210	7.9

Notes: Reinforcement assumed balanced at right angles (warp and weft). Higher reinforcement levels of aramid and carbon fibres also possible (like glass).

In comparison with aluminium alloy and steel, it can be seen that glass-reinforced plastic is limited in stiffness. This can be overcome, inter alia, by increasing the thickness, as the flexural stiffness increases as the cube of the thickness *(Table 2.8)*. Note the reduction in mass compared with steel, though not necessarily with aluminium. Other methods are listed within the text. ▶

Table **2.8**

Thickness and mass for comparable stiffness in flexure of reinforced plastic, steel and aluminium using data from Table 2.7

material	thickness (mm)	comparative mass
E glass reinforcement		
low	2.9	0.51
medium	2.4	0.48
high	1.9	0.43
other fibre reinforcement: (V_f = 30%)		
aramid	2.1	0.34
carbon	1.6	0.28
metallic alloys		
steel	1.0	1.00
aluminium alloy	1.4	0.50

Stiffness can be increased in a number of ways:

- more of the fibres can be laid in the desired direction: if reinforcement is laid in a unidirectional manner, strength and modulus are doubled in the fibre direction though they fall off sharply in the non-fibre directions *(Figure 3.8).*
- aramid or carbon can be substituted for glass with a corresponding increase in modulus. *(Table 2.7)*
- thickness can be increased *(Table 2.8).*
- sections which do not need to be flat can be curved *(Figure 3.1).*
- ribs, stiffeners or flanges can be incorporated *(Figure 3.4).*
- 'sandwich' panels with outer skins of reinforced plastic and a foamed form of the resin as a core can be used *(Figure 3.4).*

If any shear, torsion or secondary loads are present, then fibres could well be needed at appropriate angles to the main load axes *(Section 3.5).*

Other properties

It is necessary to ensure that other physical, mechanical and chemical properties can also be met. The location of relevant data is given in *Table 2.9.*

Table **2.9**

Location of data within book

aspect	reference section
fibres	5.1
variation with angle	3.3
resins	5.3
composites	5.5
smoke emission	4.9
operating temperature	3.6, 7.7
environmental effects	3.7
energy absorption	3.5
creep, fatigue	3.5
pultruded sections	5.5
strain rate	7.7

Calculations

There are computer programs available which can be used to facilitate structural calculations. Some are specifically intended for design with composites and others for structures in which the material is assumed to be isotropic *(Section 3.3)*.

The higher the cost of the raw materials, the greater the need to minimize material usage and to place the fibres in optimum amounts at optimum angles.

Scaling factors

Factors which take into account material properties are useful in establishing an embodiment design and differentiating between different materials. Some such factors are listed in *Table 2.10*.

Table **2.10**

Scaling factors in geometrical design

design consideration	scaling factor
minimum mass for given stiffness	ρ^2/E
highest fatigue life for spring	σ/E
minimum mass for pressure vessel	ρ/σ
lightest panel for given elastic stiffness	$\rho/E^{1/3}$
maximum stored energy per unit mass for a rotating member	σ/ρ

Note: values for modulus (E), strength (σ) and density (ρ) listed in Tables 2.7 and 5.1 and Figure 5.1.

For example, the choice of material for the inertial storage rotor described in *Box 1.4* depends upon the ratio of strength to density (ρ/σ). From *Table 2.7* it is clear that E glass is better than aluminium alloy or steel, and this was one reason why fibre reinforced plastic was selected.

2.9 **Function, volume and cost**

The principal cost elements comprise the raw materials and the conversion into a component (*Table 2.11*). The less material required to make a given component, the lower is the cost. Thus for reinforced plastics, care in selecting and cutting out fabrics, moulding direct to final dimensions to minimize trimming, and a low scrap rate will all help to reduce the amount of material required.

Table **2.11**

Cost elements for manufacture

cost elements

- fibre
- conversion into fabric
- resin
- tooling and plant
- labour
- trimming and waste
- quality control

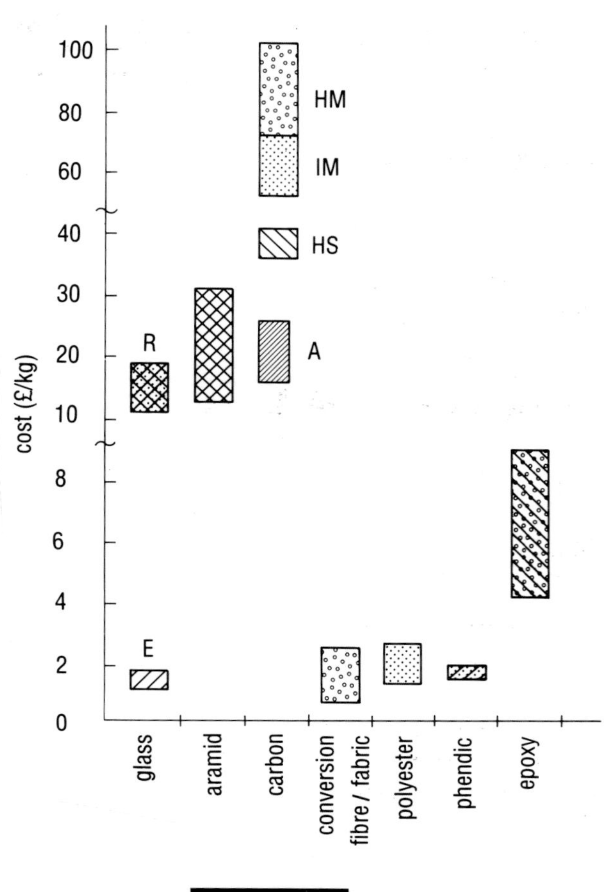

Figure **2.14**

Raw materials costs in 1992. For carbon fibres, HS refers to high strength, HM to high modulus, IM to intermediate modulus and A to non-aerospace grades. Conversion refers to processing of fibre into a fabric

The advantage of stock items is that their costs can be readily established as they are produced in high volumes and hence at low cost.

Material costs

The basic cost (list price) in tonnage quantities of the most commonly used fibres and resins is given in *Figure 2.14*. Due to an oversupply of performance fibres, prices of these fibres are volatile at present (1992). The price band for aramid reflects the difference between coarse and fine yarns (low and high cost respectively). The lowest band for carbon is for non-aerospace qualified materials and the other bands for aerospace materials with varying property levels *(Figure 5.1)*.

The price difference between the fibres is substantial – as much as 20 to 1 in the case of carbon to E glass. So the usual strategy is to design initially in glass unless the component is stiffness-limited for which aramid or carbon will have to be considered from the start (*Figure 5.1*).

The designer is therefore forced at an early stage to decide whether to design for lowest cost (as typified by glass mat/polyester), least mass (carbon/epoxy prepreg or thermoplastic sheet compound), or optimum performance (which could involve a hybrid reinforcement of glass, aramid or carbon to provide the appropriate balance of properties).

The finer the yarns, the higher the fibre cost (top of each band in *Figure 2.14*) and that of conversion into a fabric. Such materials do however tend to give the highest performance (high in *Table 2.7*).

Fabrication costs

These costs depend upon the process used to fabricate the component and the size of the production run. The chief factors for a given process are the tooling and the labour costs. In general, the higher the cost of the tooling the lower the cost of the labour.

For **low volume** production, contact moulding requires little tooling as there is only one mould surface and moulds can often be made from reinforced plastic *(Section 6.2)*. However, the labour cost per item is high.

For **larger volume** production the use of closed-mould tooling *(Section 6.4)* should be considered to reduce labour costs.

For **high volumes**, use of metal tooling and automated loading may be justified. The automotive parts in *Box 1.1* were made in this way.

One obvious advantage of working with reinforced plastic materials is the ability to make a small number of prototypes, using contact moulding and simple tooling, to check the acceptability of the product and its performance.

One can then **scale up** production using a fabrication route more suited to volume production (*Figure 2.11*). This will generally require more expensive tooling and a decision may even be made to make the component from a metal rather than a composite. Any change in production method will of course mean that the performance of the component must be rechecked.

The level of quality assurance must be appropriate to the load-bearing capability of the component and its structural task (functional, secondary or primary) and this must be considered as the design is refined *(Sections 7.10 and 7.11)*.

Production costs

The principal production costs, made up of tooling, labour, trimming and quality assurance, can only be estimated by working closely with the fabricator and material suppliers on the basis of the layout drawings. Changes that enable costs to be reduced will usually be incorporated during the detailed design phase unless they have a direct influence on the performance of the component.

Manufacture and testing of prototypes is sensible where either the material or process is used in a novel way or the performance of the component is critical to the operation of the product. Any modifications can then be incorporated during the final design phase.

2.10 Environmental impact

The design of a product affects its environmental impact in numerous ways – mode of use, product life, reuse, refurbishment and ultimately recycling. The designer, manufacturer and client need to work closely together to establish the most suitable and viable design option.

For example, some polymeric blends and fibrous reinforcements may be simpler to recycle than others. Moreover, it may be easier to separate out the constituent materials if fewer are used. Coding materials, particularly plastics, is essential for identification for recycling.

Various countries, notably Germany, have developed schemes for evaluating the environmental impact of a product in order to encourage good design practice. A label is awarded to those products which have been assessed and have minimal impact. The EC has now agreed a European scheme, the EC eco-label, which will co-exist for the time being with the various national schemes.

Eco-label award

Features of the EC eco-label award include:

- voluntary basis.
- ecological criteria to be developed for specific product groups.
- consultation with interest groups.
- a validity period of about three years.
- complement other existing or future Community labelling schemes.

A life cycle analysis involving a 'cradle to grave' approach is required, starting with the use of raw materials and culminating in their disposal.

The mark is illustrated in *Figure 4.1*, together with other marks that can be attached to products. A number of product types are currently being assessed for the European scheme and that of washing machines is discussed in *Box 2.7*.

Box **2.7**

Eco-labelling of washing machines

The assessment of the environmental impact of washing machines considered the following phases:

- production (including consumption of raw materials and manufacture of component parts).
- distribution (including manufacture of packaging).
- use (including generation of electricity used by machine and treatment of waste water).
- disposal (including transport to disposal site, energy used to extract and recycle useful materials, and the amount of solid waste produced).

The consumption of water, energy and detergent had the largest overall impact (*Figure 2.15*). Issues considered in setting the qualifying levels included:

- percentage of machines exceeding the level for a particular criterion.
- whether it was possible to distinguish between the performances of the various machines.
- whether there were significant differences in the technologies used in the different machines.
- percentage of machines achieving the levels of all key criteria.

From this survey, the ecological criteria were drafted and these included threshold levels for energy, water and detergent consumption, performance criteria, the need to identify the major plastic types using a suitable code and specific instructions to users.

These draft criteria are expected to be adopted (1992) following circulation and amendments to the draft proposal.

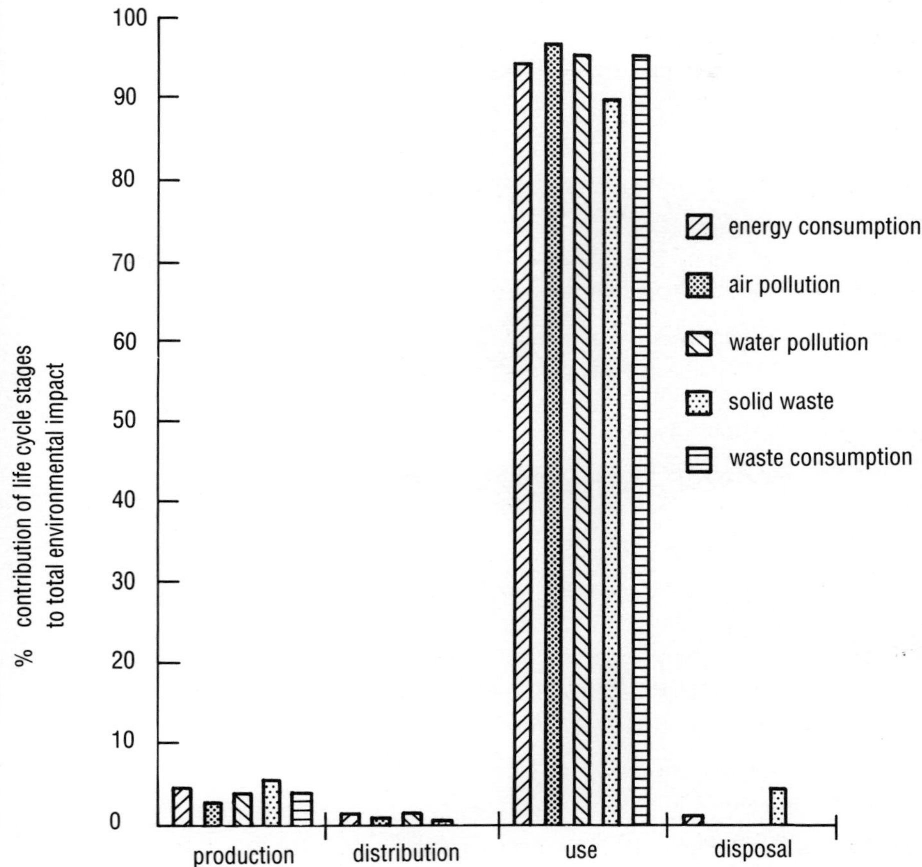

Figure **2.15**

Percentage contribution of life cycle stages to total environmental impact of washing machines (Eco-labelling Board)

2.11 **Reference information**

General references: design

Polymers and Polymer Composites in construction, L C Holloway, Thomas Telford, London, 1990 – good on design and analysis.

Design Data for Fibreglass Composites, Owens Corning Fibreglass, Ascot, 1985 – invaluable source of information.

Design with Advanced Composite Materials, L Phillips (ed), Design Council, London, 1989 – comprehensive chapters on various aspects of technology including materials, joints and applications.

Engineering Materials, M F Ashby and D H Jones, Pergamon, Oxford, 1980 – comparative properties of materials and dependence of structures on properties.

Elsevier Materials Selector, N A Waterman and M F Ashby (ed), Elsevier, Barking, 1991 – vol 3 composites – materials selection and specification.

General references: techniques

Successful Composite Techniques, K Noakes, Osprey, London, 1989 – sound practical information.

Polyester Handbook 13th ed, Scott Bader, Woolaston, 1986 – strong on practical applications of polyester resins.

2.1– 2.3 **Design**

A guide to Managing Product Design BS 7000, BSI, Milton Keynes, 1990 – essential reading and provides check lists of aspects to consider during design.

Engineering Design, G Pahl & W Beitz, Design Council, London, 1984 – sets out in a lucid manner a systematic approach to design.

An Introduction to the Design Process, K M Wallace, University of Cambridge, 1990 – design notes in Department of Engineering.

2.4 **Materials**

Design Manual Engineered Composite Profiles, J A Quinn, Fibreforce, Runcorn, 1988 – good on design and properties of pultruded profiles; load tables also available.

Engineered Materials Handbook Vol 1 Composites, ASM International, Metals Park, 1987 – compact and comprehensive source.

K Noakes *loc cit* 1989 – useful for prepreg materials.

2.5 Properties

Composite Materials Handbook, M M Schwartz, McGraw Hill, New York, 1983.

L Phillips (ed) *loc cit* 1989 – chapter 3: properties of thermoset composites and design.

Engineering Design Properties of GRP, A F Johnston, BPF, London, 1986 – invaluable source of information even though compiled in 1978.

For more detailed information consult references at end of *Chapter 5*.

2.6 Processing

Composites: a Design Guide, T Richardson, Industrial Press, New York, 1986 – useful summary of fabrication processes.

2.7 Shape and jointing

Engineering Design Process, B Hawkes and R Abinett, Longmans, Harlow, 1985 – strong on design philosophy including ergonomics.

Adhesives in Engineering Design, W A Lees, Design Council, London, 1984 – standard reference on adhesive joints.

'Stuck for good', in *Advanced Composites Engineering*, Sept 1989.

2.8 Embodiment design

L C Holloway (ed) *loc cit* 1990 – design and analysis of composite structures.

K Noakes *loc cit* 1989 – design studies using composites.

Materials Optimizer, Fulmer, Stoke Poges, 1980 – includes detailed list of scaling factors in geometrical design.

2.9 Costs

J A Quinn *loc cit* 1988 – useful cost parameters for fabrication.

2.10 Eco-labelling

Community Eco-label Award Scheme, EC regulation 880/92, Official Journal of the European Community 11/4/92.

Environmental Labelling Fact Sheets 1 and 2, Eco-labelling Board, London, 1992.

Environmental Labelling of Washing Machines, H E Durrant et al, PA Technology, Melbourn, 1991 – Pilot Study for DTI/DoE.

General sources of information

Materials Information Service, Design Council, London – runs helpline and information service +44 071 839 8000.

Plastics and Rubber Advisory Service, 5-6 Bath Place, Rivington Street, London EC2A 3JE +44 0839 506 070 – general advice – refer to *Section 1*.

Composites Processing Association, Tannery Court, Westover View, Crewkerne, Somerset TA18 7AY +44 0460 72 8 70 – information on processing and processes.

RAPRA Technology Ltd, Shawbury, Shrewsbury, Shropshire +44 0939 250383 – general advice on material selection and fabrication.

Production Engineering Research Association (PERA), Staveley Lodge, Melton Mowbray, Leicestershire LE13 0PB +44 0664 501 501 – general advice on fabrication technology.

Materials or Engineering Divisions within National Laboratories including:

National Physical Laboratory, Queens Road, Teddington, Middlesex TW11 0LW +44 081 977 3222.

National Engineering Laboratory, East Kilbride, Glasgow G75 0QU +44 03552 20 222.

AEA Technology, Harwell, Didcot, Oxon OX11 0RA +44 0235 821 111.

UK Eco-labelling Board, Eastbury House, 30-34 Albert Embankment, London SE1 7TL +44 071 820 1199 – advice on eco-labelling schemes and criteria.

Range of stock items

A brief list of what is available is set out in *Tables 2.12* to *2.14*, compiled from various manufacturers' catalogues. Suppliers can be identified through the trade associations given above.

Table **2.12**

Typical range of pultruded sections (dimensions in mm)

type	range (mm)	type	range (mm)
solid rod		**tube**	
diameter	2 – 50	diameter	6 – 150
		wall	2 – 4
box		**bar**	
width	25 – 100	width	6 – 150
wall	2 – 6	wall	6 – 25
angle		**channel**	
width	10 – 150	height	14 – 70
wall	3 – 12	width	25 – 250
		wall	2 – 12
I beam		**wide flange beam**	
height	100 – 300	height	75 – 200
width	50 – 150	width	75 – 200
wall	6 – 12	wall	6 – 9
studding and nuts			
diameter	9 – 25 (UNC)		

Table **2.13**

Typical range of available glass reinforced flat sheet

thickness minimum (mm)	maximum	width (m)	maximum length (m)
0.6	25.0	1.2	2.4
1.0	3.0	3.0	up to 60.0

Note: only thin panels can be curved.

Table **2.14**

Other stock items

	diameter (mm)	**wall thickness** (mm)
filament-wound tubes	29 – 58	2 – 14
	thickness (mm)	**mass per m²** (kg/m²)
sandwich panels	14 – 52	3 – 8

Detailed design considerations

chapter 3

summary

In detailing the design, further consideration must be given to aspects which differentiate reinforced plastics from other types of materials. The extra freedom that the materials allow the designer, in terms of shape and form, has to be balanced by the need to ensure that there is sufficient reinforcement present to withstand loads. Other properties can be tailored to the needs of the design by varying the nature and type of reinforcement.

Introduction

The stage has now been reached in the design where the following aspects of the design have been established (Table 3.1):

Table 3.1

Inputs to detailed design

design inputs
- embodiment design completed
- fabrication route identified
- material combination chosen
- budget cost established
- environment impact considered
- relevance of codes and standards determined
- level of fire resistance evaluated

Of these inputs, all but the last two have been considered in *Chapter 2* and these are discussed in *Chapter 4*.

If it is necessary to check the embodiment design, whether to gauge client and customer acceptance or to validate aspects of the design, this should be done before spending much time on detailing. If changes have to be made then it is cost effective to make them now.

During this stage, each of the steps in the embodiment design *(Figure 2.3)* is repeated in some depth. If alterations are made to either the materials or the processing route then the effect on the preceding aspects of the design needs to be checked.

Where information is lacking, it will often be necessary to validate concepts by undertaking appropriate tests *(Chapter 7)*. Test pieces or components should always be made using both the desired materials and production route.

3.1 Shape, size and finish

Shape

This has a major effect on the choice of process and form of reinforcement *(Section 5.2)*. If the component is a straight replacement of an existing component then the geometry could be fixed and the manufacturing process selected to suit the geometry *(Section 6.1)*.

If there is some freedom in determining the geometrical shape then there is scope for optimizing the design through rationalising the type of reinforcement or the processing route (novel suspension, *Section 8.3*).

If space is available then full use can be made of the ability of reinforced plastics to take up complex curvature to provide both additional stiffness and a pleasing aesthetic design *(Figure 3.1)*.

Size

As the size of component affects the choice of manufacturing process (*Figure 3.2* and *Section 6.1*), the appropriate use of joints can reduce the component size to provide a wider choice of process.

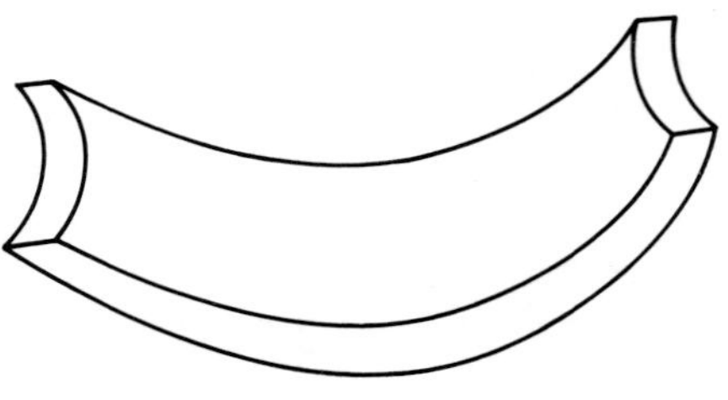

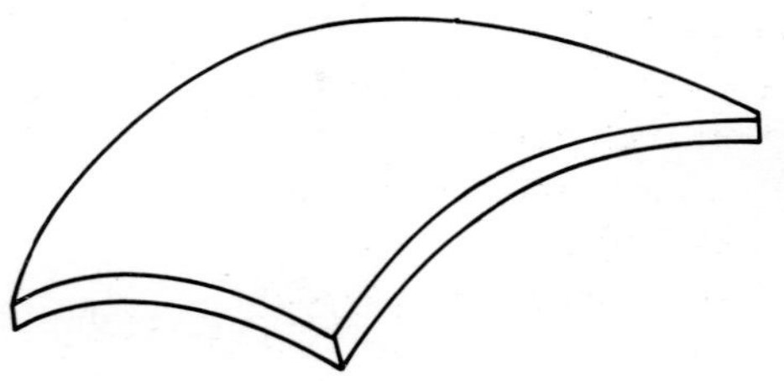

Figure 3.1

Examples of complex curvature possible with reinforced plastics

If joints can be permitted then the component could be made of two or more parts. These parts may be fabricated by either the same or different processes. In the example shown in *Figure 3.3*, the joint lines have been inserted so that one set of components has cylindrical symmetry and the other axial symmetry, so different manufacturing processes can be used.

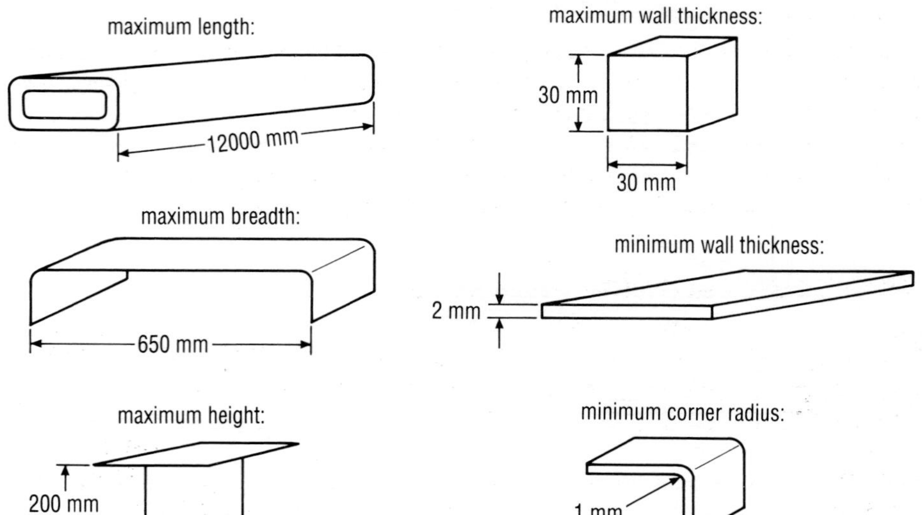

Figure 3.2

Typical restraints on size and shape for pultruded products (after Fibreline)

Thickness

Both thick and thin sections can pose problems in manufacture:

- sections less than about 2 mm need to have good surface layers which are fibre free.
- sections greater than about 20 mm cause problems because of possible process limitations or high exothermic temperatures which arise during curing of the resin as a result of poor conductivity.

The pultrusion sections in *Figure 3.2*, for example, permit a wall thickness variation of 2 to 30 mm.

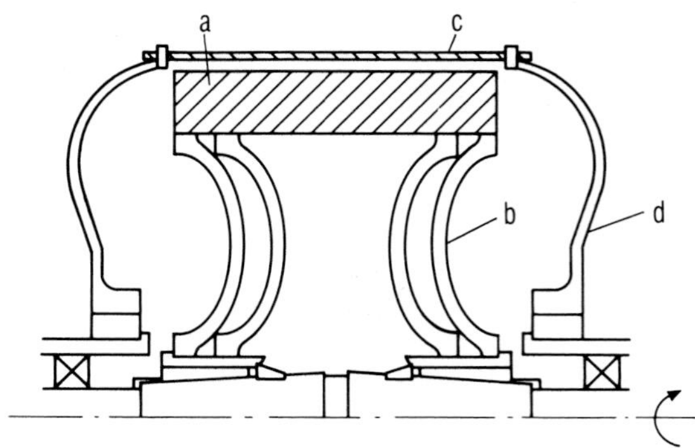

Figure **3.3**

A flywheel inside its safety casing. The flywheel rim (a) and liner (c) have cylindrical symmetry and so can be filament wound while the flywheel hubs (b) and end plates (d) are well suited to being moulded (Medlicott, Society of Automotive Engineers)

Foam cores. Stiffness can be substantially increased in thin sections by using a low density foam or honeycombe core between the laminate skins (*Figure 3.4a*). A range of core materials are available with varying densities (typically 100 to 200 kg/m^3) and varying mechanical properties such as shear strength, compression and shear modulus.

Impact performance is important in applications where slamming loads can occur as bond continuity must be maintained across the skins and core. Processing therefore needs careful consideration to lay up the core section to follow the geometrical shape and to select and apply a suitable adhesive.

Variations in thickness are possible provided that there is a way of getting the fibres to fill all the available space: for example, the ribs in *Figure 3.4b* are easy to form with a suitable moulding compound like DMC or SMC *(Section 5.4)*, but much more difficult with fabric. One method of filling variable cross-sections is to use a thick piled fabric like needle mat *(Section 5.2)* with a typical compressibility of 4 to 1.

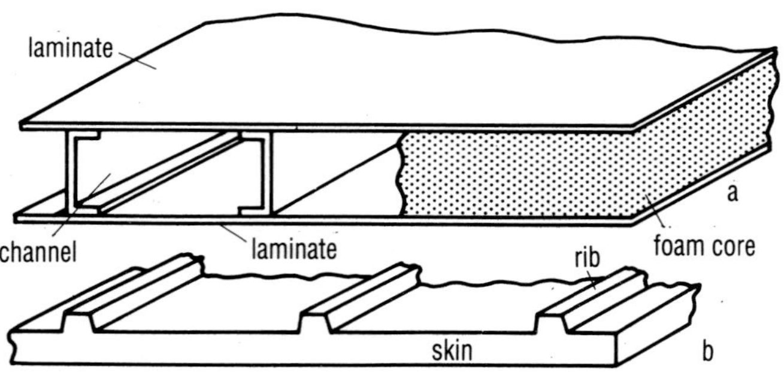

Figure **3.4**

Methods of stiffening thin sections: (a) Sandwich panel of two laminates using either a foam core or stiffeners which could be pultrusions (b) Section stiffened by use of ribs; note that thickness of rib should be similar to the thickness of the skin

Minimum radii depend upon the manufacturing technique (example in *Figure 3.2*) and the requirement to ensure that fibres can penetrate into the corners to provide some resistance against impact and wear. The more generous the radii, the easier it is to ensure that a fabric (if used) can be draped without having to be cut and tailored. Guidance is normally available from the moulder.

Dimensional tolerances

These can be achieved without machining if attention is paid to both design and manufacture. Machining needs to be avoided because it will remove the protective surface layer, take away any self colouring (if within the gel coat) and expose the underlying fibres. To achieve specific dimensions, allowance must be made for shrinkage during curing of the resin *(Section 5.3)*. This will depend on a number of process factors and also on the proportion of fibre by volume, as the fibres prevent the resin from shrinking. This is an aspect on which the designer has to work closely with both the moulder and the material supplier. The larger the component, the larger the total shrinkage, and again consideration should be given to the possibility of joining two smaller components. An example of a large moulding is the front end of the Central Line train on the London Underground, which is about 2.5 m in diameter.

Shrinkage can be minimized by one or more of the following steps:

- through curing *(Table 5.7)* in the mould or on the mandrel.
- making the mould slightly oversize to compensate for shrinkage.
- use of a low shrinkage resin *(Table 5.10).*
- use of suitable fillers and additives *(Table 6.4)*; some moulding compounds are formulated on this basis *(Box 5.8).*
- increasing the proportion of fibre to resin, by using a fabric rather than a mat, for example *(Figure 5.2).*

Surface finish

Variations in fibre concentration can lead to differential shrinkage and even to cracking, so fibre-deficient areas need to be avoided.

The finish is also affected by the nature of the processing route. The use of gel coats to provide a resin-rich surface is ideal but is only possible with some production processes *(Tables 6.2 and 6.3).*

The following options are available to achieve a hard-wearing surface:

- the addition of a surface veil or fabric, either of a glass or a non-glass type.
- the use of an intrinsically hard resin such as phenolic resin, as in brake linings.
- an additional surface layer, which may be sacrificial in type and renewed periodically if needed; polyurethane-based materials are often used for this purpose, as for example in the leading edges of wind turbine blades.
- the addition of hard particle fillers in the surface (or gel coat) layer.

3.2 Load paths and reinforcement selection

Having defined the shape envelope as fully as possible, the next step is to check that the component can withstand the design loads, both internally and externally.

If these loads are not known then they should be measured or calculated rather than assumed, because with reinforced plastics it is possible to vary the strength to match the principal loads.

Load paths

Internal loads, such as those encountered with rotating machinery or vessels and tanks, can usually be clearly defined. External loads are more difficult as they may require consideration among other aspects of how the load is transferred across a joint.
If different materials are used on either side of a joint then their respective material properties must be considered. For example, differential thermal expansions could lead to the joint getting tighter or looser.

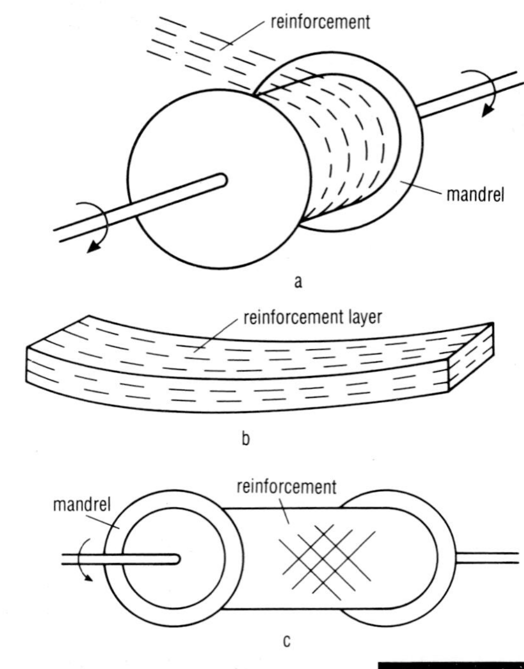

Figure 3.5

(a) Hoop winding of filaments or tapes to balance radial stresses in a cylinder. (b) Moulding of a unidirectional fabric for a leaf spring in which the prime loads are transverse to the surface of the spring. (c) Winding fibres or fabrics at 45° to axis to withstand torsional loads in a tube

It is a good design rule to to preserve material type across any joint which is either highly stressed relative to the material's capabilities or critical to the integrity of the structure. Material type can then be changed at some less highly stressed position.

Fibre lay-up

The basic method is to orientate the fibres in the direction of the principal stresses.This can be achieved by emplacing rovings, either individually or more likely in the form of fabrics or prepregs *(Section 5.2)*. This is particularly effective for simple shapes and loadings where it is generally possible to orientate the fibres to maximize their load carrying capacity *(Figures 3.5 and 8.3)*.

For more complex loadings, the designer needs to work closely with the fabricator and the material suppliers as more ingenuity is required. Options include:

- winding tows at specific angles.
- laying up woven or knitted fabrics in specific orientations; fabrics can also be designed with suitable drape and fibre orientation *(Section 5.2)*.
- laying up mats with random tows (chopped strand or continuous filament) *(Section 5.2)*.
- short length fibres which can orientate in any direction and are usually in the form of compounds *(Box 5.8)*.

These options give rise to different design solutions as they affect the key design parameters *(Box 3.1)*.

Box 3.1

Design options for low mass and low cost

Mass versus cost

The designer can trade off mass against cost, as the higher the fibre volume fraction and geometric alignment, the better the properties and the lower the mass, though generally at a higher cost (Table 3.2).

Table 3.2

Design solutions to minimize mass or cost

reinforcement	option	geometric efficiency	volume fraction
unidirectional filaments or fabric	low mass high cost	1.0	up to 65% up to 55%
bidirectional fabric	medium mass medium cost	0.5	30 – 45%
short length fibres	high mass low cost	0.4	25 – 35%

Geometric efficiency is a measure of the directionality of the fibres. The value is 1.0 if all the fibres lie in one direction and 0 if none do. Two examples are shown in *Figures 3.6* and *3.7*.

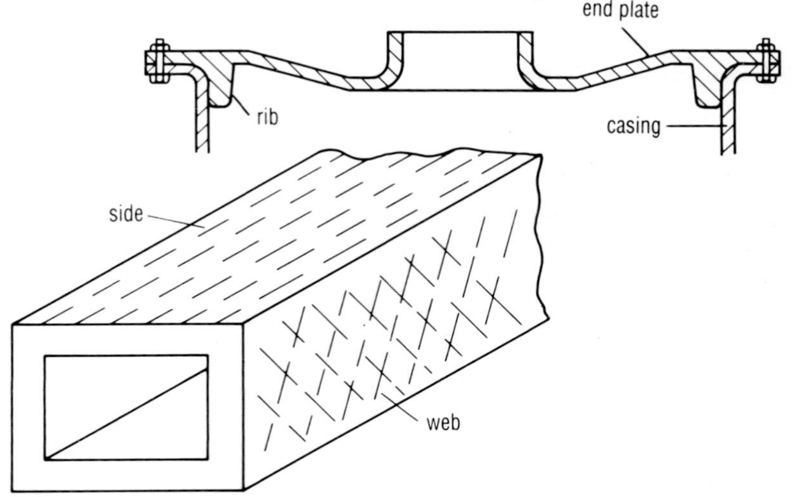

Figure 3.6

Casing plate in which the major loadings are isotropic and perpendicular to the plane of the component (Medlicott, Society of Automotive Engineers)

Figure 3.7

Structural beam in which the major loadings are longitudinal and in bending. For minimum mass, fibres should be laid longitudinally along sides and at 45° along webs

The casing plate is a low-cost option using a needled fabric which conforms to the shape without tailoring on laying up. Note that the rib which provides both extra stiffening and location for the mating part is made using a foam-core insert. To maximize the mass of the structural beam, fibres should be laid longitudinally along the sides and at 45° along the webs, which can be achieved using prepreg fabric.

Fibre length

For high-strength structures, continuous fibres are essential because they ensure continuity in the load path down to a microscopic scale and because such fibres (or fabrics) can be located in a reproducible manner in every component. For lower-strength composites, discontinuous fibres will suffice provided that two criteria are satisfied:

- no large areas are fibre-deficient.
- the load can be transferred efficiently across the resin/fibre interface.

Fibre type

There is a variation in strength of up to two for various 'alloys' or heat treatments of each of the main reinforcing fibres *(Figure 5.1)*. So additional strength can be obtained by:

- substituting the next stronger fibre with no mass penalty, although at extra cost.
- changing fibre type.

Mixtures of fibres can be used to improve the balance of properties through having:

- different layers with different fibres or fabrics.
- mixed fibre types within a layer by using a 'hybrid' fabric. This option is not straightforward and needs to be considered with care.

Resin type

The initial choice of resin has been based on the following considerations: application, fabrication process, operating temperature, cost, properties and fire resistance. The choice can be modified as the design is refined though it is more likely to involve a different blend of the same base resin. This type of information needs to be obtained from the supplier.

3.3 Strength and stiffness

Both strength and stiffness depend upon the properties of the fibres as well as those of the matrix, the latter becoming more important in directions in which the fibre fraction is low. The rule of mixtures *(Box 3.2)* works well for predicting the modulus for all types of fibres and fibre contents. Prediction of strength is more difficult by this simple rule because the stress-strain curve *(Figure 5.8)* is not necessarily linear to failure.

Both strength and stiffness need to be calculated for each of the primary stressed directions and for any other non-fibre direction in which there may be an appreciable load as the strength and stiffness fall off rapidly with increasing angle from the fibre direction *(Figure 3.8)*.

Box **3.2**

Properties of composites

Rule of mixtures

Properties p depend upon the respective contribution of the fibres p_f and matrix p_m, the fibre volume fraction V_f and the geometric efficiency g_e *(Table 3.2)*. The rule of mixtures assumes that the contributions of the components can be summed linearly in proportion to their respective fractions for a unidirectional composite:

$$p = p_f g_e V_f + p_m (1 - V_f)$$

When p_f, g_e and V_f are large then the fibres will dominate the resultant property, and conversely the matrix will dominate if p_f or V_f are low.

In many instances the rule does provide a basis for the design if the relevant input data is available, but this should be checked against real data before finalising the design.

If greater accuracy is required or the rule is not applicable, then there are numerous texts which describe how to undertake such calculations.The problem can become complex if finite element analysis is required so there is some virtue in keeping the design 'simple'.

Value of testing

One cannot assume that the same type of fibre and resin will always give the same properties, as resins and fibres do differ in their ability to transfer load.

Tests should therefore be carried out on a particular fibre/resin combination to check some of the fundamental properties such as stiffness, strength and strain to failure in the principal fibre directions *(Chapter 7)*. Loading conditions should be appropriate to the application because properties may not be the same in tension, flexure, shear or compression *(Section 5.12)*.

If these results are lower than those predicted by the rule of mixtures, then possible reasons are:

- the fibre has inappropriate finish or size for resin.
- the resin has not been appropriately cured *(Tables 5.6* and *5.7)*.
- the processing was incorrect.

A low level of properties is less likely to occur if moulding compounds or prepregs *(Section 5.4)* are used, as the fibre, resin and processing are all standardized. Moreover, data should be available on which to base the design. If doubt exists, the material supplier should be consulted at an early stage in the design.

Variations in strength and stiffness with angle are greatest for unidirectionally (UD) aligned fibres and smallest for a random mat (like chopped strand) and intermediate for the woven roving fabric (WR) with a balanced number of fibre tows in the warp and weft directions *(Figure 3.8)*.

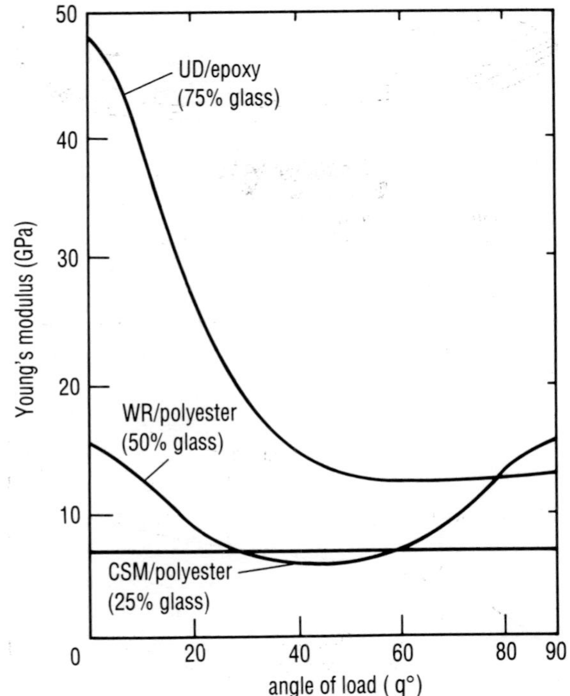

Figure **3.8**

Variation in Young's modulus in tension with the angle of load relative to the principal fibre direction for different types of reinforcement in GRP. A similar variation with angle applies to the tensile strength (Johnson)

Calculation

Conventional structural formulae can be used if:

- the fibre distribution is isotropic in three dimensions (low fibre volume fraction).
- there is uniform packing of fabric layers and quasi-isotropy within the layers (orthotropic).

A useful analysis of this is contained within *Design procedures for plastic panels* (A F Johnson and G D Sims, 1987) which sets out in a systematic way how to consider various types of loading on panels which are clamped in various ways. Ribbing can also be allowed within this analysis.

Computer-aided laminate analysis has been developed which enables material lay-up to be optimized, although this could involve a cost penalty on fabrication. Some other computer programs are also available, primarily from the aerospace industry, to help structural applications *(Section 3.10)*.

Design procedures are prescribed in certain codes of practice and standards such as BS 4994 and BS 7159 *(Section 4.7)*, and these provide guidance for other types of calculation.

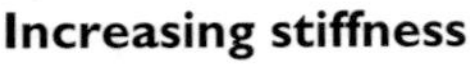

Increasing stiffness

As designs with composites tend to be limited by stiffness rather than strength (unlike metals), the various options include:

- using stiffer fibres by substituting some or all the glass for carbon fibres *(Section 5.1)*.
- modifying the geometry or using a sandwich type construction *(Section 3.1)*.
- increasing the fibre fraction in the property-limited direction *(Tables 2.7 and 3.2)*.

Since the resin must be able to transfer the load to the fibre, there is an optimum fibre volume fraction for each type of reinforcement and above this level properties can decrease.

The criterion for adequate stiffness is usually set by the design itself, and calculations are facilitated by the composite's linear response to load. Strength criteria are more complex because they depend both upon the design and the way in which damage accumulates in the materials.

3.4 Damage accumulation, tolerance and failure

The very nature of the materials comprising a composite means that the damage process is different to that in metals. The designer who understands the way in which damage accumulates in these materials should be able to use this to advantage in designing a failure mode that can be benign. One way of achieving benign failure is to lay up the fibres (or fabric) in such a way that the failure occurs within the matrix rather than by breaking fibres.

Figure 3.9

Schematic failure of a leaf spring between the laminate planes

For example, in the case of a vehicle leaf spring made from GRP, a loss of stiffness and failure would occur progressively due to delamination between the laminate planes *(Figure 3.9)*. Nevertheless the spring would still have load-carrying capability and so the vehicle could be driven to a garage.

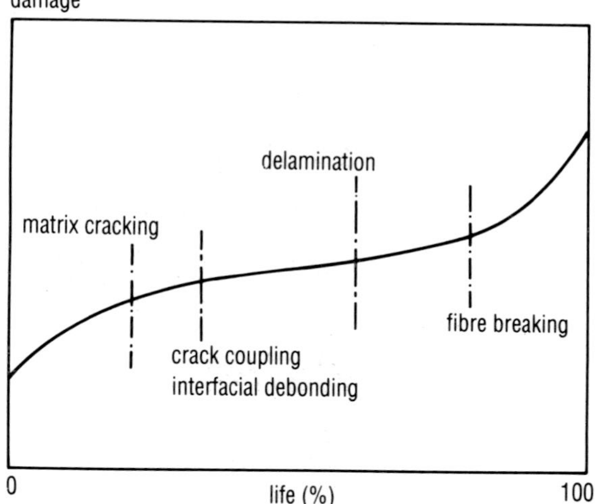

Figure 3.10

Schematic of the damage modes as a function of service life. An analogous pattern would be observed if the sample was simply loaded to failure in a short-term test (Reifsnider)

Damage accumulation

During service, damage will accumulate at load levels and in patterns that are controlled by the properties and geometry of the constituent materials and the manner in which they have been processed.

At low stress levels, damage tends to accumulate in the matrix by microscopic cracking, fibre debonding and delamination of the laminate plies over the whole of the stressed volume. As these tend to be widely isolated, damage can accumulate with few outward signs.

At higher stress levels, fibres impart a measure of toughness through their ability to deflect cracks or to absorb energy by fracturing or pulling out. Failure finally occurs either through delamination, as discussed in connection with the leaf spring, or through fibre failure such that the load can no longer be carried.

Because the damage can occur in different forms and be widely dispersed *(Figure 3.10)*, no single method of measuring the amount of damage will suffice *(Section 7.11)*. This makes inspection difficult, so designers should incorporate benign modes of failure wherever possible.

Damage tolerance

The varied nature of the damage enables some composite materials (eg GRP) to be classed as damage tolerant. One aspect of this tolerance is their ability to sustain damage due to a sudden excessive load and yet may be able to continue to perform their normal function for a short time until they can be inspected.

Another aspect is their lack of notch sensitivity, since a crack will not necessarily grow from a notch because of the dispersed nature of the damage. This can apply even under cyclic loading in corrosive environments; additional damage will only accumulate if new fibres are exposed to the environment. As with other design parameters, it is possible to enhance damage tolerance through selected lay-up of the laminate plies or by selecting tougher fibres, fabrics or resins.

Inspection

Since catastrophic failure is generally unacceptable for structural components, the designer must specify intervals at which the component must be inspected. In design terms, these can be arrived at by observing the rate at which damage accumulates and the major mechanical properties decrease with time *(Figure 3.11)*.

Failure criteria

The three stages of damage are identified in *Table 3.3* and illustrated in *Figure 3.11*.

The transition from stage 2 to 3 is critical in terms of safe design. If it cannot be detected by monitoring one or other property, then the design must fail benignly or have a service life which is less than the time of the onset of failure (stage 3).

A decrease in stiffness of a set amount, say 10%, could be one failure criterion. Since many designs are stiffness limited such a decrease may be observable, as for example in the vibration spectrum of a rotating machine or the additional deflection of a structure.

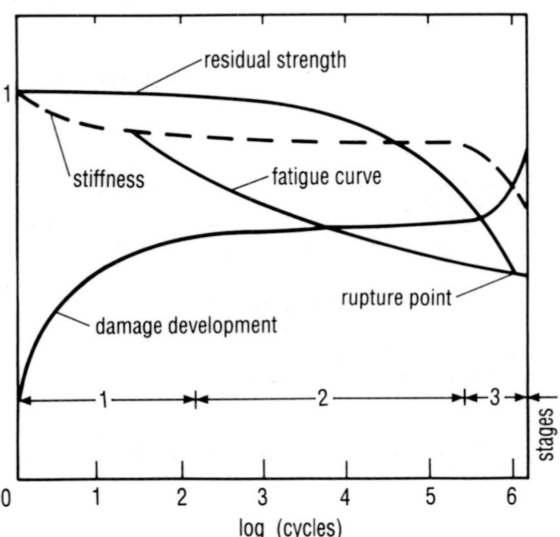

Figure **3.11**

Decrease in stiffness and strength during fatigue as a function of damage accumulation and life. This damage sequence seems to occur for many types of loading (Reifsnider)

Table **3.3**

Stages of formation of damage during fatigue (after Reifsnider)

stage		duration	changes
1	initial	short	small drop in stiffness
2	secondary	variable	little change
3	tertiary	short	ever-increasing leading to failure

Other criteria might be strength limited. An indication of damage on a local scale may be easier to detect, such as cracking of the gel coat.

3.5 **Loading modes**

Unless the fibres are orientated uniformly in all directions (ie short length fibres), the resulting composite will have non-isotropic properties *(Table 5.12)* and so the designer has to consider how to cope with various loading modes within the design.

Composites are generally stronger in tension than in compression because fibres close to the surface can buckle. This can be checked by testing coupons in both tension and compression. An indication can be obtained from a flexural test *(Figure 3.12)* as visible damage will occur on whichever is the weaker surface (tension or compression).

Compression damage can be mitigated by ensuring good bonding between the fibre and resin and that the fibres are not too close to the surface by using a good gel coat or surface fabric *(Section 3.1)*.

Interlaminar shear forces can result in delamination because of the low shear strength between reinforcement layers. Incorporation of three dimensional fabrics or other third axis reinforcement such as needling can provide an increase in strength. In-plane shear loads can be carried by reinforcement layers at ±45 degrees. The designer may also select a resin with a high shear strength or alter the design to reduce the shear forces. This can also be used to advantage if a benign failure mode is required as in the example of the leaf spring *(Figure 3.9)*.

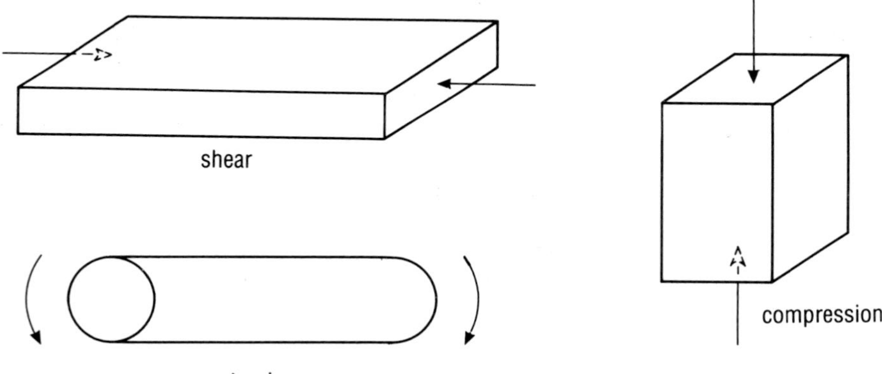

Figure **3.12**

Forces acting on structural elements

Torsion can usually be overcome by ensuring that fibres are present at the appropriate angles *(Figure 3.5)* using either a suitable tape or winding process.

Matrix properties tend to dominate in directions transverse to the fibres. Thus for unidirectional fibres or fabrics, some reinforcement may also be needed in other directions.

In general, fibres should be apportioned according to the principal loads, but if this is not possible then an increase in mass may have to be suffered in order to ensure sufficient stiffness or strength. The use of dedicated fabrics is an elegant method of reinforcement, but alternatives include using a mixture of long and short length fibres or the use of emplacement techniques such as fibre lay-up, tape laying or preforming *(Section 6.4)*.

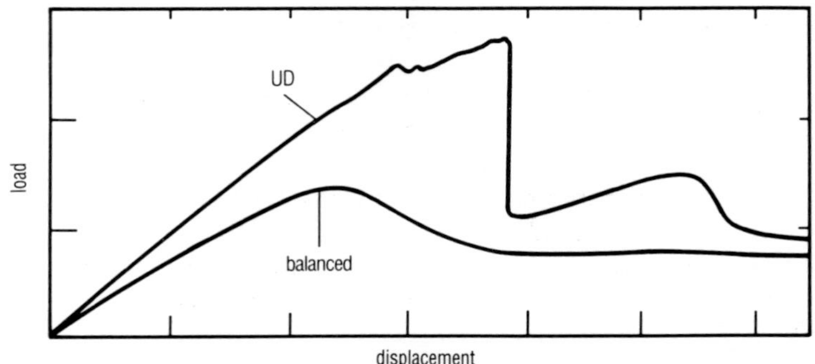

Figure **3.13**

Response of two different laminates to loading – one with the fibres fully aligned (UD fabric) along the load axis, the other being a balanced woven roving (WR fabric); normal strain rate (Sims)

Toughness and impact

The toughness of composites arises from the ability of the fibres to blunt incipient cracks and the work necessary either to fracture them or pull them out of the matrix. This varies with fibre lay-up and fabric, as can be observed in the response of such materials to load *(Figure 3.13)* in which the area under the curve represents the ability to absorb impact energy.

Glass fibre reinforced plastics tend to become both stiffer and stronger at high strain rates *(Figure 7.18)*, such as those that occur for example in impact situations. While this gives a higher margin of safety, the material (and therefore structure) is also tougher *(Figure 3.14)*.

Since glass fibres are tougher than carbon, and aramid in turn tougher than glass, hybrid fabrics can be used to provide a balance between stiffness and toughness. Toughened resins can also improve resistance to crack propagation *(Section 5.3)*.

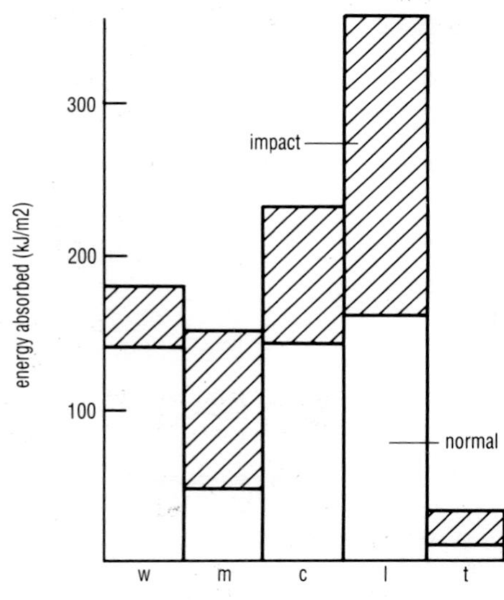

Figure **3.14**

Energy absorption under normal and impact loading for various types of reinforcement:
w – fine woven fabric;
m – random mat;
c – cross-ply fabric;
l – UD fabric along fibre direction;
t–UD fabric in transverse direction
(Sims)

Long term properties

The design has so far concentrated on achieving adequate response to short-term loading. Clearly, over the service life of the design, damage accumulation, tolerance and life expectancy will be a function both of the past and expected load history, and operating temperature.

Long-term loading can comprise any combination of sustained load (creep) or fluctuating load (fatigue). These properties tend to be fibre-controlled provided that there is adequate bonding between the fibres and the resin. In addition, the maximum service temperature must be kept at least 20°C below the heat distortion temperature of the resin (BS 7159).

This aspect can therefore be regarded as a design parameter, since varying the fibre type and volume fraction will alter the material's resistance to such loadings.

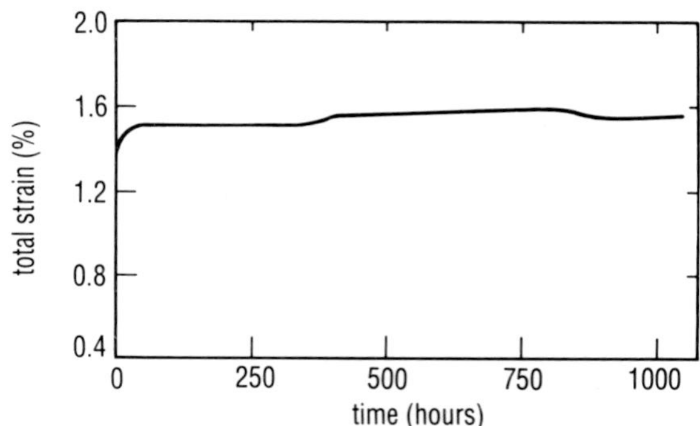

Figure **3.15**

Change in strain in UD aramid (Kevlar 49)/epoxy subjected to 80% of its ultimate tensile strength at room temperature (du Pont)

Creep

Creep deformation can occur under load at elevated temperatures, so resins should be selected with heat distortion temperature at least 20°C above the maximum service temperature to minimize dimensional changes.

If a component will be subjected to prolonged loadings at elevated temperatures then creep measurements should be made to characterize the selected material and process route. By establishing creep rates at some temperatures and loads, it may be possible to predict the effect at others.

Aramid fibres are particularly good in this regard *(Figure 3.15)* and have therefore been used in applications such as pressure vessels and flywheel rotors.

Fatigue

Glass reinforced plastics have good resistance to fluctuating loads, their fatigue strength decreasing in a monotonic fashion out to one million cycles and beyond for a variety of matrices *(Figure 3.16)*.

The data in *Figure 3.16* has been normalized by dividing the (maximum) fatigue strength by the ultimate strength. This provides a framework for assessing variations in materials, loadings and test conditions.

Carbon reinforced plastics hardly degrade under fatigue loading and display a fatigue limit which is close to the lower bound of their static strength. Aramid fibres have a resistance intermediate between carbon and glass *(Box 3.3)*. It is uncertain whether at low strains so little damage accumulates that a fatigue limit exists *(Figure 3.17)*.

Figure **3.16**

Normalized tensile fatigue of long fibre reinforced thermosets and short fibre reinforced thermoplastics (Mayer)

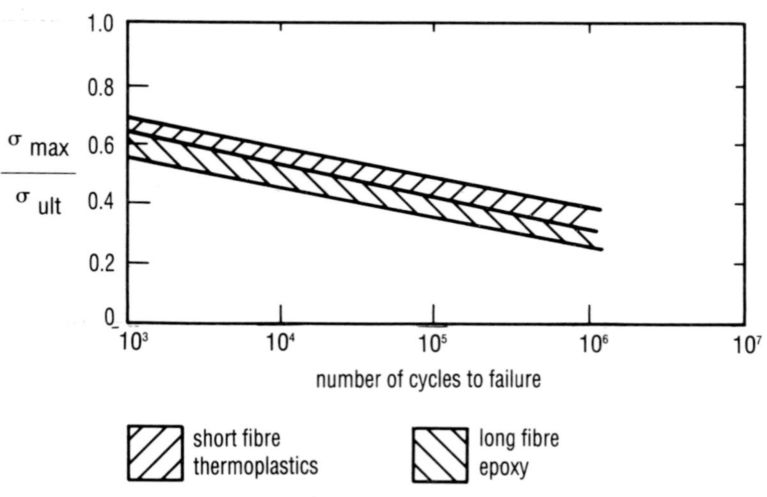

To achieve good fatigue life the following practices are recommended:

- as with static strengths, the higher the working stress the greater the need to incorporate a higher fibre volume fraction or fibre orientation in the direction of the principal fatigue loads.
- substitute some aramid or carbon fibres (or fabrics) if the necessary strength cannot be obtained with glass reinforcement alone.
- testing should be carried out to ensure that the material combination and process route produces the required level of properties *(Section 7.8)*.
- the effect of scale not being well known, tests should be also be carried out on full-size components before the design is put into service *(Section 7.10)*.

Box **3.3**

Fatigue resistance

Fibre type

The existing data for three principal reinforcing fibres show a definite trend *(Figure 3.17)*. Fibres having the highest stiffness (carbon) are the most resilient under fatigue loading while the fibres with the lowest stiffness (glass) show the greatest change.

This suggests that strain is an important parameter. So if fatigue were the only design parameter, then there is a clear choice between glass, aramid and carbon.

There appears to be a limiting strain below which a very long life can be achieved. At present, a limit of 0.1 of the matrix strain to failure (about 0.3% total for many resins) has been proposed by at least one certification society for glass reinforced plastics. This can be contrasted with the 0.2% strain quoted in BS 4994.

Framework

This has been developed to cover the various aspects of loading, testing and component geometry. While this has only been verified in detail for a specific material (balanced weave glass/epoxy), it has also been applied successfully to other materials and loadings. This covers:

- test frequency – may be as high as 50 Hz provided that any temperature rise is minimal and less than 10°C.
- test temperature – the upper limit should be 20°C (or more) below the heat distortion temperature of the resin to comply with code recommendations *(Section 4.4)*.
- specimen pre-treatment in water – may include immersion and heating to 100°C.
- notches and holes.
- damage tolerance – by pre-loading to a stress higher than the fatigue stress.

This results in a single stress/cycle (S/N) curve in which all the data can be represented by normalizing the fatigue strength by the ultimate strength of the samples measured at the loading rate of the test (analogous to *Figure 3.16*).

Figure **3.17**

Maximum fatigue strain as a function of cycles for various woven fabric reinforcements in epoxy resin (Fernando)

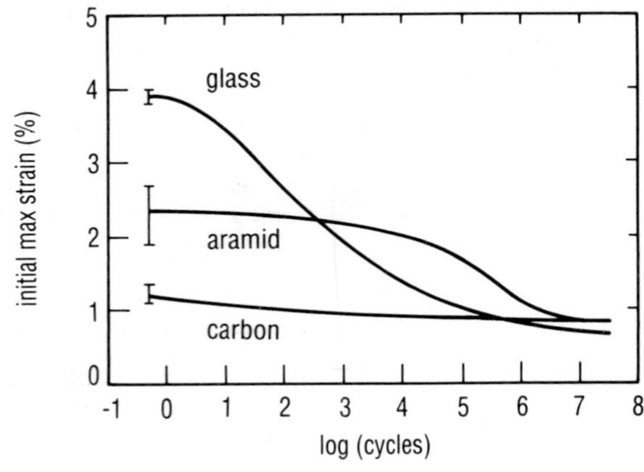

3.6 **Effect of temperature**

Resins generally have lower melting or softening points than fibres *(Sections 5.1 and 5.3).* It is therefore the properties of the resin rather than those of the fibres that degrade as the temperature increases. Consequently the resin is less able to transfer load to the fibres and they in turn are less capable of restraining the resin from deforming under the applied load, so the load-carrying capability of the composite drops. As with creep, the higher the heat distortion temperature of the resin, the better the temperature capability of the composite *(Figures 3.18 and Table 5.10).*

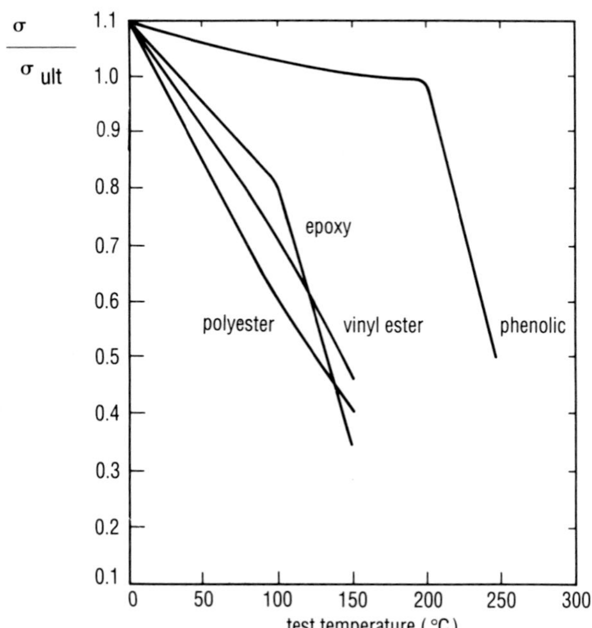

Figure **3.18**

Flexural strength of similar glass reinforced composites with various matrix resins (strength normalized to 25°C) (Forsdyke)

Thermoplastic and aramid fibres may also lose strength and stiffness at elevated temperatures if their softening points are similar to the resin.

Phenolics have the highest working temperature of the common resins, but recent needs within the aerospace industry have led to the formulation of both thermoset and thermoplastic resins which can be used up to 300°C *(Section 5.3).*

At low temperatures (below freezing), strength and stiffness do not generally drop and may even increase in the fibre direction *(Section 7.7).* Because of the freeze/thaw cycle, it is desirable to ensure that there is good fibre/resin bonding, that the resin is fully cured *(Section 5.3)* to remove any remaining volatiles, and to check the performance of any component which undergoes a large amount of temperature cycling.

The effects of prolonged temperature exposure (creep) have already been discussed *(Section 3.5).*

3.7 **Environmental effects**

There are many environmental influences which reinforced plastics may be subjected to *(Table 3.4).*

Table **3.4**

Environmental effects

- temperature
- ultra-violet radiation
- erosive particles
- lightning
- humidity
- sea water
- salt spray
- corrosive liquids

All these environments can damage the surface of a reinforced plastic component, and if the surface layer is penetrated by cracks or diffusion, the underlying fibres (and resins) will also be exposed and may be degraded.

Surface protection

Good design should include one or more of the following, depending upon the severity of the exposure:

- a good quality surface layer with no porosity and no reinforcing fibres within the surface.
- a resin rich layer (gel coat), particularly for polyester resins for which specific resistant coatings are often available.
- incorporation of a layer of surface tissue to provide a thicker and more durable surface – for example C type glass fibres which have good corrosion resistance *(Table 5.1)*.
- the addition of a layer on top of the surface with appropriate properties to mitigate any attack – this could even be sacrificial in nature and so could be renewed as necessary – for example, a layer of polyurethane resin is often applied on the leading edges of wing or propeller blades.
- the insertion of a metallic mesh in the surface layer to act as an electrical path to mitigate lightning strikes.
- the provision of an inert liner (often thermoplastic) which prevents the surrounding environment from coming into contact with the body of the material.
- the selection of a matrix resin with properties which will provide additional protection to the reinforcing fibres if the surface layer is penetrated.

Chemical resistance

Resistance to chemical attack tends to be resin-dependent *(Figure 3.19)*, so resin manufacturers should be consulted about the best approach for a particular application.

Some reinforcing fibres are also more resistant than others and this provides an additional design option *(Table 5.2)*. For example, glass is inert while carbon will oxidize above 200°C. On the other hand, carbon is more inert than glass to many corrosive liquids.

Some harsh environments where composites have been successfully used are summarized in *Table 3.5*.

Testing

Testing must often be carried out to check the application *(Section 7.7)*. Some standard test gear is available like weatherometers for ultra-violet, humidity and salt spray. Immersion is also used and one such test for pipes is shown in *Figure 4.2*. Use is made of both temperature and chemical concentration to accelerate tests, but it may be difficult to extrapolate to other environmental conditions.

Table **3.5**

Some environments in which composites are used

environment	typical component
hot and wet	aircraft engine components
chemical inertness	pipes for effluent and corrosive liquids
sea water resistance	boat hulls
weatherability	roof tiles

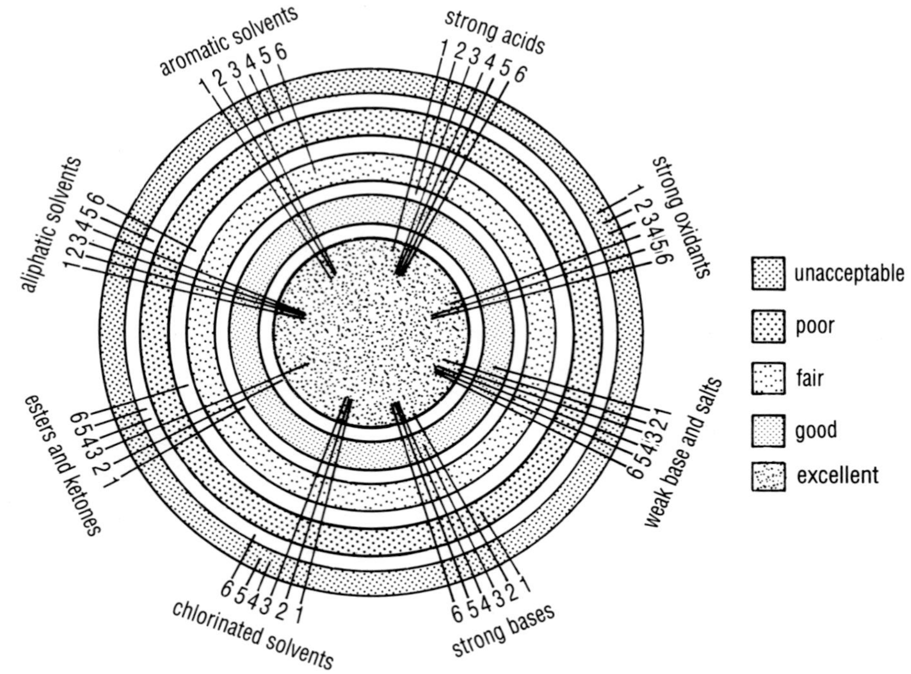

Figure 3.19

Chemical resistance at 23°C
1. glass fibre polyester
2. glass fibre/epoxy
3. furane
4. polyvinyl chloride
5. polyethylene
6. polypropylene
(Design data, OCF)

3.8 **Joints**

Joints in reinforced plastics need careful design and aspects are discussed throughout the text *(Table 3.6)*.

One can summarize good jointing practice as follows:

- make correct allowance for shrinkage during moulding and exercise particular care if mating components are made by different processes or contain considerably different fibre fractions *(Figure 3.20)*.
- mould direct to final shape and size so that only trimming is required.
- joint surfaces should not be machined thereby exposing fibres (and incurring cost).
- ensure mating surfaces retain shape during any post-curing by restraining surface movement if necessary.
- ensure that reinforcement in the vicinity of bolt holes is adequate to cope with strains induced on tightening bolts and do not drill too close to edges.
- coat drilled holes with resin to reseal exposed fibres – this is particularly important in harsh environments.
- use washers with bolts and oversize if there is any danger of cracking around the holes on tightening.
- at joints with dissimilar materials use wavy and/or spring washers with bolts to take out any effects of differential thermal expansion.

Testing

It is sound practice to check the ability of the joint to withstand both normal and abnormal service loads. Quality assurance and quality control are particularly important for adhesive joints, where it very difficult to check the joint quality in a non-destructive manner.

Table **3.6**

Design considerations of joints

aspect	design consideraton	reference (section)
type	separable or fixed	2.7
number	advantages	3.1
surface	quality of finish	3.1
location	relation to material type and stress	3.2

Mechanical joints

For mechanical joints, tests must be carried out to check whether the joint has adequate strength. These include:

- bolts – tighten the bolt until its thread is stripped. If the material is damaged by this process, would it be visible? If one relies on the washer to spread the load, what would happen if the washer were inadvertently left out?
- inserts – carry out a similar test. Does the insert remain in place under applied load?
- creep relaxation – apply a load in excess of what would normally be encountered and check the torque necessary to retighten the bolt after set intervals of time. Retightening once is generally acceptable.

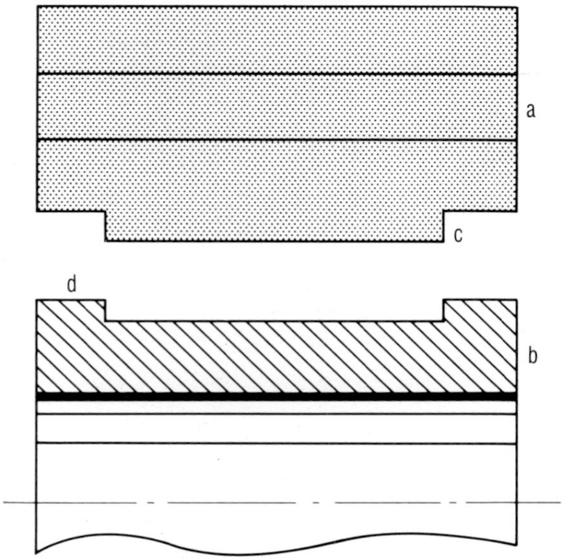

Figure **3.20**

Typical glued joint. Component (a) has been filament wound and the step (c) has been achieved by having a raised step on each edge of the mandrel. Component (b) has been moulded and the mating step (d) created by a recessed step in the mould. Both joints were created without need for any further machining

Adhesive joints

Adhesive joints require more consideration than mechanical ones. The joint should be designed in such a way that it carries a compressive element within its load pattern. Adhesive joints are not suitable for tensile loading and peel and cleavage loads should be minimized *(Figure 2.13)*.

Adequate surface preparation is essential. For example, traces of mould release agent must be removed – usually light abrasion and solvent wiping is sufficient, but this may not be true for some thermoplastics which are not cleaned by common solvents. Care must also be taken to apply the adhesive in such a way that air is not trapped within the joint.

The main types of structural adhesives have been described *(Box 2.5)* and toughened adhesives give good performance under impact loading.

Welding is possible with thermoplastics and such techniques are available for use on a range of materials.

Repairs may often be possible using adhesive jointing. Repair kits for motor vehicles have been available for many years, generally for repairing corroded metal parts, and they seem to have been very effective. More recently, similar techniques have been developed for aircraft due to the increased use of composites. Metallic aircraft panels are now being successfully repaired with composite material patches.

3.9 **Recycling and reclamation**

Thermoplastics

These materials can be reset by heating. This means they can be reused in a different shape or form, provided that the various components can be separated, and the fibre content is suitable. A further possibility would be to granulate the materials and reuse them to form alloys with other thermoplastics.

Thermosets

The chemical, thermal and mechanical stability of thermoset composites – which makes them materials of choice for many applications – also represents the challenge to recycling. Once cured, these materials cannot subsequently be softened by further heating, and so cannot be reused in their existing form. Structural components may be separated into a form in which they can be directly reused as rods, panels and tubes, though the variety of chemical types and reinforcements may make the process uneconomic.

If not recycled, there are three possible disposal routes:

- mechanically crushing and using as land fill, although the resistance to biodegradation is a disadvantage.
- grinding into large granules for use as hard core for roads or buildings (also in polymer concrete).
- incineration as a fuel, which would require monitoring of possible long-term toxic emissions, and their effects on the environment. A pilot plant, established in France in 1990, has demonstrated this to be a feasible method, with minimal atmospheric emissions. A full-scale facility would require an annual consumption of approximately 10,000 tons to ensure economic viability.

Pyrolysis provides a chemical method of recycling thermosets – the composite is decomposed by heat to recover organic monomers which can be used in the synthesis of new products and new polymers. This process, well researched in the USA, enables various organic gas products (such as methane, propane and propylene) to be recovered in significant quantities.

The complete recycling of composite materials has become the focus of considerable research in recent years. New recycling legislation in several European countries to ensure environmental protection has necessitated this work.

It is possible to treat discarded thermoset products as a raw material for use in new thermoset or thermoplastic composites, without incineration or other chemical preparation. Controlled mechanical granulation of cured, styrene crossed-linked polyester sheet moulding compound (SMC) gives a blend of free glass fibre, ground polyester powder, and calcium carbonate filler. This blend can be used as a filler and reinforcement in polyester bulk moulding compound (BMC) and in polyolefin thermoplastic moulding compounds.

Engineering properties have been measured on thermoset and thermoplastic composite sheets which have been compression moulded from various formulations based on granulated SMC. BMC made with recycled SMC maintains 70% or more of the mechanical properties of a standard BMC formulation. In polyolefins containing recycled SMC, significant improvement in modulus is obtained at equivalent strength. This demonstrates recycled SMC to be a valuable raw material that contributes to the physical properties, and can be used in the design and production of new composite parts.

3.10 **Reference information**

General references

Composite Materials Series, B Pipes (ed), Elsevier, Amsterdam, 1988 – six volumes covering aspects of the technology.

Handbook of Composites, A Kelly and Yu N Rabotonov, Elsevier, Amsterdam, 1985 – four volumes covering many aspects of composite materials science and technology.

Specific references

3.1 **Shape**

Polymers and Polymer Composites in Construction, L C Holloway (ed), Thomas Telford, London, 1990 – architectural use and aesthetics of composite structures.

Successful Composite Techniques, K Noakes, Osprey, London, 1992 – sound advice for sandwich panel construction using honeycombe core.

'Sandwiches for strength', *Advanced Composites Engineering,* March 1989 – provides examples where core panels have been successfully used.

'Core materials and trends in the boating industry', *IRPI,* June 1989 – describes range of foam core materials available.

3.2 **Fibre lay-up**

Design Data for Fibreglass Composites, OCF, Ascot, 1985 – useful design information clearly set out.

'Aramid, carbon and glass fibre reinforcement materials for composites' R Kleinholz and G Molinier in *Fibreworld* No 22 p13, 1986.

3.3 **Strength and stiffness**

L C Holloway (ed) *loc cit* 1990 – design and analysis of composite structures.

Design Procedures for Plastic Panels, A F Johnson and G D Sims, NPL, Teddington, 1987.

'Rethinking the design process', K Edwards in *Advanced Composites Engineering,* September 1989.

BS 4994 and BS 7159 – methodology and safety factors as applied to pressure vessels and pipes (see *Section 4.5*).

3.4 **Damage accumulation**

'Damage-tolerant polymer composite systems', K L Reifsnider in *Journal of the Institute of Metals* p50, November 1988.

3.5 **Properties**

Engineering Design Properties of GRP, A F Johnston, BPF, London 1986 – good review of properties though large scatter bands on much of the data.

Kevlar Data Manual, du Pont, Wilmington, 1986.

'Understanding Charpy impact testing of composite materials', G D Sims in *Proc 6th Int Conf on Comp. Mats* p3494, 1988.

'Fatigue behaviour of hybrid composites', G Fernando et al in *Journal of Materials Science* 23 p3732, 1988.

'Effect of test conditions on the fatigue strength of a glass-fabric laminate', G D Sims and D G Gladman in *Plastics & Rubber* p41 May 1978 and p122 August 1978.

3.6-3.7 **Environmental effects/temperature**

'Phenolic matrix resins – the way to safer reinforced plastics', K L Forsdyke in *43rd Ann Conf, Soc of the Plastics Ind,* 1988.

Recommendation for creep temperature limit comes from *BS 7159 'Code of practice for chemical plants or sites',* BSI 1989.

A F Johnson *loc cit* 1986 – creep measurements (to 1978).

L C Holloway (ed) *loc cit* 1990 – Chapter 7: end-use performance and time-dependent characteristics.

Most information is contained in specialized conference proceedings on either composites or chemical plant such as those published by IMechE (London) and BPF.

'Practical aspects of applying lightning protection to aircraft and space vehicles' K G Payne et al in *Proceedings of 8th International SAMPE Conference*, La Baule, France, 1987.

3.8 **Joints**

Adhesives and the Engineer, W A Lees, MEP, London, 1989 – review of role of adhesives in structural and mechanical engineering industries – for earlier book refer to references to *Chapter 2.*

L C Holloway (ed) *loc cit* 1990 – comprehensive introduction to adhesive and bolted joints and repair methods.

'Repairing the damage', K Armstrong in *Advanced Composites Engineering,* p48 Winter 1987.

3.9 **Recycling**

Various initiatives are under way either by associations such as RAPRA Technology (Shrewsbury, England) and IKF (Aachen, Germany) or manufacturers of thermoplastics.

'BASF has developed a new process for GRP recycling', *Fibreglass Facts* No 31, 1989 – discusses reuse of reinforced thermoset materials for use in bulk moulding compounds.

Recycling SMC, R B Jutte and W D Graham OCF, Toledo, 1991.

'Plastics materials recycling, zero waste concept' C Desnost in *Composites*, 3 177/8, 1991.

'Car builders urged to reuse mountains of scrap vehicles' *The Times*, London, 21.8..92.

Computational aids

Analysis of laminated composite behaviour can be considerably simplified by using personal computer programs. These are useful tools for persons with some experience in analysing composites although not always transparent to users. Possible limitations include the type of reinforcement lay up and the selection of a suitable failure criterion.

A series of 14 codes dealing with failure analysis, frequency effects and buckling problems is available from Engineering Science Data Unit, 27 Corsham Street, London N1 6UA +44 071 490 5151

Other programs include:
Compcal (University of Delaware, Newark, Delaware, USA)
Mic-Mac, RANK and LAMP (Think Composites Software Users Club,Dayton, Ohio, USA)
Lamanal (Severn Consultants, Newport Pagnell, UK)
Coala (College of Aeronautics, Cranfield, UK)
Nolan (AEA Engineering, UK)
Laminate (AEA Industrial Technology, UK)
Class (Materials Science Corporation, USA)

For further information, contact:
Richard Lee, AEA Technology, Harwell, Oxon OX11 0RA, UK +44 0235 821 111

Safety, standards and codes

chapter 4

summary

The formation of the single European market will permit the free movement of goods, services, people and capital throughout the European Community and EFTA countries. Product liability, safety, regulations, codes and standards are considered and the way in which these apply to a product or structure. One aspect considered in some detail is that of fire safety.

Introduction

Products designed with reinforced plastics are no different to others in that the designer must consider existing legislation and consult relevant codes and standards. These are changing rapidly owing to the growth of consumer protection and the extension of markets beyond national boundaries (such as the formation of the single European market).
The design sequence, set out in *Figure 2.1*, indicates three aspects to be considered:

- legislation: national, international and now European
- guidance provided by codes and standards
- general issues of product safety and liability

The working of the single market is discussed first. In the following sections, directives, codes and standards which bear on design in general and on design with reinforced plastics in particular are outlined. One aspect, fire safety, is considered in more depth.

4.1 Single Market

The single market to be created in Europe from 1 January 1993 forms an area without internal frontiers in which the free movement of goods, persons, services and capital is ensured. In order to achieve this aim, some 280 measures have been agreed or are under consideration by the Council of Ministers of member states of the EC. These are diverse in character and affect many aspects of product design, ranging from construction to consumer safety.

The majority of the single market measures are directives which require either legislation or promulgation for these to become law in each country; a smaller proportion are regulations or decisions.

Box **4.1**

Single Market and economic integration
The European Community is based on the Treaty of Rome, which was signed in 1957 by France, Germany, Luxembourg, Belgium, Netherlands and Italy. More recent members are United Kingdom, Ireland, Denmark, Spain, Portugal and Greece. The original treaty has been amended notably in 1985 to allow the formation of the single market by 1993 and in 1992 at Maastricht and Oporto.

Article 2 of the Maastricht agreement affirms that 'the community shall have as its task . . . the promotion of a harmonious and balanced development of economic activities, sustainable and non-inflationary growth respecting the environment, a high degree of convergence of economic performance, a high level of employment and of social protection and the raising of the standard of living and quality of life, and economic and social cohesion and solidarity amongst member states'.

The Maastricht treaty, entitled 'The Treaty on European Union' cannot come into effect until it has been ratified by all member states. The European Union spans both the Community and the complex web of new institutional relationships which is now developing between the member states.

In Oporto, it was agreed to form the European Economic Area (EEA) between the EC and EFTA countries (Austria, Finland, Iceland, Liechtenstein, Norway, Sweden and Switzerland). This agreement, once ratified and adopted in law, will extend the single market principles and regulations to these countries; other areas of cooperation include education, research and development and the environment.

The EEA will create a market of 375 million consumers accounting for more than 40% of the world trade, and is scheduled to come into force soon after the completion of the single market (1 January 1993).

At the same time, the EC is also establishing collaborative links with Eastern Europe and Turkey, which are likely in the course of time to lead to a further extension of the EEA.

This drive towards economic integration should be viewed as stimulating rather than hindering trade with other major trading blocks such as Eastern Europe, North America and Asia/Pacific by creating a market with common standards and harmonized regulations.

For example, a recent directive adopted by the EC 'confirms the interest of an international standardization system capable of producing standards that are actually used by all the partners in international trade and of meeting the requirements of Community policy'.

Directives

The EC directives are concise documents specifying the essential requirements necessary to achieve their aims. In many instances, supplementary agreements are required to interpret the provisions of such directives such as the eco-labelling of products (*Section 2.10*). Where the formation of codes or standards is required (like Eurocodes, *Section 4.6*), the EC has mandated certain European bodies to undertake this work (*Box 4.3*).

Essential requirements

In setting the 'new approach' directives, the Commission has considered its responsibility both to its citizens and to industry. So it has adopted the fundamental principle of differentiating between essential requirements and applications.

The essential requirements cover safety, public procurement and guidelines for the harmonization of standards. These are mandatory. Standards therefore need to reflect these requirements and to act as a service to industry in opening up markets.

There are several ways in which these requirements can be met:

- follow standards and codes where relevant
- demonstrate a performance comparable with the essential criteria
- use in-house standards comparable with the essential requirements

In this book, the major directives which affect design are briefly considered.

4.2 **Product liability**

People who are injured or whose property is damaged by defective products have the right to sue for damages, and product liability is the term given to laws affecting such rights. EC legislation has removed the need to prove negligence and action can be taken against the producer, raw material supplier, processor, importer or someone who brands a product as his own (*Box 4.2*).

Box **4.2**

Product liability

Product liability and safety provisions

The EC directive on product liability was adopted in 1985 and came into force in the UK civil law via the Consumer Protection Act of 1987. The directive lays down the principle of strict product liability regardless of fault.

With the passing of this act, anyone injured by a defective product, whether or not the product was sold to them, can sue a supplier without proof of negligence. A wide range of products are covered including component and raw material suppliers. Although a design consultant cannot be sued directly by the consumer for a mistake in a design which causes a product to be defective, the consultant is still liable for his actions to the designer or producer (though not under the Consumer Protection Act).

Defective product

When deciding whether a product is defective, a court will take into account all the relevant circumstances, including:

- the manner in which a product was marketed
- any instructions or warnings that are given with it
- what might reasonably be expected to be done with it
- the date the producer supplied the product

Possible defences include:

- the state of scientific and technical knowledge at the time of supply of the product was not such that a producer of products of the same description might be expected to have discovered the defect if it had existed in his products whilst they were under his control
- the defect was a necessary consequence of complying with the law
- the defect was not in the product at the time it was supplied.

Advice to designers

There are numerous instances in which a design has been found to be inadequate and has led to defective products: for example, the location of a fuel tank within a vehicle chassis, the lack of adequate quality control on manufacture or the interchangeability of parts on assembly.

The inappropriate selection and use of materials can also lead to injury. During the fire in London's Kings Cross Underground Station, for example, it was not the origin of the fire but the rapid spread via other materials (of an organic nature) that prevented people from escaping.

It is therefore important to adopt the best working practices. The implementation of BS 7000 and the methodology set out in Chapter 2 of this book will assist in the general design process. This task is simplified if suitable standards and codes already exist, as this indicates that a consensus has been reached.

With reinforced plastics, the technology is still evolving, so the relevant codes may not be available. The designer then has to consider working practices in related areas and seek advice from designers and engineers involved in such activities.

Material suppliers and fabricators are often able to assist in the selection of materials, and to say whether anyone has attempted a similar use. Design or engineering consultancies should also be considered where appropriate. Special vigilance is needed, however, with materials like reinforced plastics for two main reasons:

- their properties only arise as the result of manufacture, so any error or incorrect procedure could result in a product whose properties differ from those on which the design was based.

- products may be assembled with insufficient care if the workforce is not acquainted with the procedures appropriate to such materials, for example trying to drive a bolt through two holes that do not overlap.

It is clear that the more highly stressed a component or a hazardous environment may be, the more care needs to be taken in its design. For load bearing structures, the certification societies such as Lloyds Register or Det Norske Veritas (DNV) should also be consulted at an early stage. This is not only to obtain suitable advice, but also because potential insurers might well insist on their approval as a condition of insurance.

4.3 Health and safety

The requirements of health and safety at work (such as the UK Act of 1974) place additional statutory responsibilities on designers which need to be considered where there is a potential risk to health. These would include:

- that an item or structure is designed and constructed so it is safe and without risks to health when properly set, used or maintained, so far as is reasonably practicable.
- carrying out or arranging for such testing and inspection as may be necessary for the performance of the proposed duty.
- ensuring that there is adequate documentation to describe the conditions of use, with updates as relevant new knowledge becomes available.

Where safety is a consideration, inspection intervals should be set in accordance with what is known about the material combination, its structural design, its loading and its durability.

For novel applications of reinforced plastics, there should be short intervals and comprehensive inspections. For recognized applications, intervals can be longer. As confidence is built up in the design and the application, such periods can then be extended.

Product safety

The EC has agreed a new directive on general product safety which will apply to all products 'intended for consumers or likely to by used by consumers supplied whether for consideration or not in the course of a commercial transaction and whether new, used or reconditioned'. It is due to come into force on 29 June 1994.

Specific requirements on producers include:

- placing only safe products on the market: 'goods subjected to the new approach directives being exempted from the general safety requirement'.
- adopting measures commensurate with the risks that these products might present, such as providing relevant information to enable consumers to assess the risks inherent in the product.

A product must be safe 'under normal or reasonably foreseeable conditions of use' and must present no risks or 'the minimum risks compatible with the product's use', consistent with a high level of protection for the safety and health of persons taking into account:

- the characteristics of the product . . . .
- the effect on other products where it is reasonably foreseeable that it will be used with other products.
- the presentation of the product, labelling and instructions for use and disposal . . . .
- categories of consumers at serious risk when using the product, in particular children.

To ensure that products comply with the general safety requirement, member states have the necessary powers to check and monitor safety and if necessary to prohibit the sale and even to withdraw a dangerous product.

For conditions under which products are deemed safe it is advisable to consult the directive. The directive is broader in scope than Part II of the UK Consumer Act of 1987, which excludes certain product sectors and second-hand goods.

4.4 **Standards and codes**

A standard or code is a guide to current working practices. It will comprise one or more of the following elements, which form the basis of the design process:

- characteristics of a material or product
- design procedures
- methods for the evaluation of properties
- limiting values or tolerance bands of certain properties

The importance of codes and standards has grown in recent years as product safety has become enshrined in law, markets have become international and materials are loaded more effectively through optimizing the design of components (*Box 4.3*). Codes of practice or design codes set out procedures, which in turn call up standards and test methods.

Box **4.3**

Standards and their organisation

Formation of standards

Standards are formulated on the basis of voluntary agreements by committees representing interested persons and bodies and are organized on either a national or international basis.

There are a very large number of national standards and there is no reason why these should not be used to comply with existing legislation provided the principle of mutual recognition is accepted in countries where the product is to be sold. The value of international standards such as those set by ISO is that there is mutual acceptance within those countries who are members (see below).

For example, once a European standard has been agreed, the member bodies have to modify their national standards to conform with the agreed standard. In 1992, some 3000 European standards are available in final or draft form and this will rise to 9000 by 1995, many associated with the use of EC directives.

Within Europe, three bodies are concerned with standardization:

- European Committee for Standardization (CEN)
- European Committee for Electro-Technical Standardisation (CENELEC)
- European Telecommunications Standards Institute (ETSI)

At present the members of CEN/CENELEC are the national standards (or electro-technical) organizations of EEC and EFTA countries, and they are represented on the technical committees where the standards are drafted.

Other organizations involved with standards which are cited in this text include:

BSI	–	British Standards Institution	(BS)
DIN	–	Deutsches Institut für Normung	(DIN)
AFNOR	–	Association Français de Normalisation	(AFNOR)
ASTM	–	American Society for Testing Materials	(ASTM)
JIS	–	Japanese Industrial Standards	(JIS)
ISO	–	International Organisation for Standardisation	(ISO)
CAA	–	Civil Aviation Authority	(UK)

Impact of the single market

As noted in *Section 4.1*, the aim of the single market is for products to be designed and manufactured to the same standards used throughout the European Economic Area.

In certain market sectors, where member states accept mutual recognition of standards and technical regulations, the status quo is legally binding following a decision in 1979 by the European Court of Justice ('Cassis de Dijon').

In a limited number of other sectors, where this cannot be applied because of reasons of public health, safety or security, the EC has issued a number of directives and these are gradually being extended (*Table 4.1*).

Table **4.1**
New approach directives of EC

scope	number	date of entry
simple pressure vessels	87/404	31.12.91
toy safety	88/378	1. 1.90
construction products	89/106	27.12.91
electromagnetic compatibility	89/336	1. 1.92
machinery	91/368	1. 1.93
personal protective equipment	89/686	1. 7.92
non-automatic weighing machines	90/384	1. 1.93
gas appliances	90/396	1. 1.92
active implantable medical devices	90/385	1. 1.93
telecommunications terminal equipment	91/263	6.11.92
equipment for use in explosive atmospheres	91/516	1. 7.93
hot water boilers	90/368	1. 1.94

** under discussion*

A third group of sectors will not be covered by such directives as there are no legal obstacles. However, the Commission will encourage standardization to remove de facto obstacles. The water, energy and transport sectors have so far been excluded from any direct regulatory initiative. However, the EC is in the final stages of implementing a set of public procurement directives which will allow contractors from other member states to bid on equal terms for public works, including those of privatized utilities.

Adoption of standards

The drive towards liberalizing trade within and between major trading blocks is leading towards the increased formulation of international standards. These can originate from existing national standards which have widespread international recognition, like the British standard on quality control (BS 5750).

In-house standards are often set by large industrial manufacturers such as Boeing or Airbus Industrie for aircraft, Ford Motor Company for cars, and agencies such as NASA or ESA for space, or major users such as London Underground Ltd for mass transit.

There is an appropriate time to originate standards. Too soon and insufficient experience could lead to inappropriate recommendations; too late and there may be difficulty in reconciling existing (de facto) standards. This timing is particularly difficult for technologies that are still evolving, like reinforced plastics.

4.5 **Quality systems**

There are now two general quality management systems in place: design, development, production, installation and servicing (ISO 9000/ EN 29000/ BS 5750); and environmental management systems (BS 7750).

The first system provides a comprehensive set of general checks from specification through manufacture to installation and testing (*Box 4.4*). An example of how such a quality assurance specification is evolved is given in *Box 4.5*.

4.5 **Quality systems**

The second system covers the impact on the environment of both the manufacturing process and the product through an examination of the whole organization. The key elements include:

- setting environmental objectives to meet legal and voluntary requirements sought by the organization to meet the needs of interested parties
- evaluating and recording the effects on the environment
- ensuring that the process and product comply with the objectives

This system requires an audit and a regular review procedure; it therefore complements the EC regulation on eco-auditing. In addition, an analysis of the product life cycle is a prerequisite for a product to comply with the eco-labelling scheme (*Section 2.10*). If assistance is required in devising such procedures, independent advice should be sought from consultancies specialising in such work.

Box **4.4**

Quality standard ISO 9000/9001

ISO 9000/9001

This standard applies to all materials and processes and is a system which covers the purchaser and the supplier. It is assessed by an independent body.

System requirements have to be established in relation to the extent of the assurance required and the functional capability of the producer. Separate parts of the standard cover design, specification and testing; manufacture and testing and final inspection and test.

The **design** function requires that the specification is formally documented and that this document and any subsequent revisions are then reviewed and agreed by both the supplier and the customer. This is facilitated by breaking the design down into its constituent elements.

Specifications may incorporate an agreed standard and/or an in-house standard. Whatever the basis of the specification, the purchaser must ensure that it will cover his or her needs, even if this requires additional testing. Such tests are generally required to ensure that the product will perform satisfactorily once fabricated from raw materials.

If changes to the specification are proposed by the supplier, then the purchaser should put the onus on the supplier to provide adequate notice before making the change. This allows the designer or fabricator to check the effects the proposed changes will have on the product.

Traceability requires that constituent materials are purchased from suppliers who are assessed for maintaining quality and that there is adequate stock control within the manufacturing plant.

Manufacture requires procedures defining the methods to be used, the equipment and the criteria for workmanship. Operators form an important part of the process by applying their own local process control wherever possible. Inspection and testing of manufactured parts ensures that these conform to the agreed specification. ▶

Tests exist for many properties. Where there are alternative test methods, the test method best suited to the application should be chosen *(Chapter 7)*. If no suitable test exists, then one must be devised. This is particularly likely to occur when components or structures are tested.

Records of calibration, quality tests, audits and customer complaints must be kept. The steps taken to analyse faults and prevent their occurence must also be recorded.

Box **4.5**

Quality control of wing blades for wind turbines

Requirement

To devise a draft system of quality assurance based on ISO 9000 which will provide the framework in which the fabricator and the client can agree the manufacturing route and the quality of the resulting product. The system will need to be applicable to fibre reinforced materials and processes.

Fabricator

The following requirements are considered to be essential:

- materials are to be covered by a certificate of conformity issued by the supplier listing the standards (national, international or in-house) to which the material has been manufactured
- the manufacturing procedure is to be specified by the fabricator and agreed by the customer
- the manufacturer is to keep a record of the materials used (type, drum number etc), and of any quality tests which have been carried out and their results
- appropriate tests to check the attainment of the requisite property levels.

Methods and procedures should be approved by an independent consultant and either a national test centre or a certifying authority such as DNV or Germanischer Lloyd.

Quality assurance tests

The following tests are to be carried out to characterize materials, processes and products (Tables 7.14, 7.19):

Table **4.2**

QA testing and recommended test methods

property	test	method
process control	monitoring cure cycle	–
	visual inspection	–
Barcol hardness	degree of cure of resin	EN 59
materials	test panel	ISO 1268
volume fraction	glass fibre	ISO 1172
	carbon fibre	ASTM D 3171
porosity	Archimedes method	ASTM D 2734
modulus, ultimate	tension or	ISO 3268
strength and	compression or	ISO 604
failure strain	flexure	ISO 178

Note that:
- tests are required of the fabricator to check their raw materials and what they have manufactured
- tests are required by the customer to check what has been received
- other test methods can be substituted if necessary

This type of quality assurance is simply to ensure that what is fabricated conforms to the designer's specification. The fitness for purpose would have to be established by additional tests where necessary (*Chapter 7*).

Product marks

Generally speaking, products which meet the essential requirements of the relevant EC directive shall carry a CE mark (*Figure 4.1*). This mark is intended to help to remove trade barriers and to allow goods to move freely across national boundaries. The procedures under which such marks can be used are still being devised, as they rely on the availability of interpretive documents, harmonized standards or a European Technical Approval.

Products manufactured under a quality manufacturing system conforming to ISO 9000/ ISO 9001 and to the relevant national or international standard for that product may carry a mark such as NF (France), DIN (Germany), BSI Kitemark (UK) or the CEN CENCER mark.

Other product identification marks include the eco-label (*Section 2.10*) and an energy label.

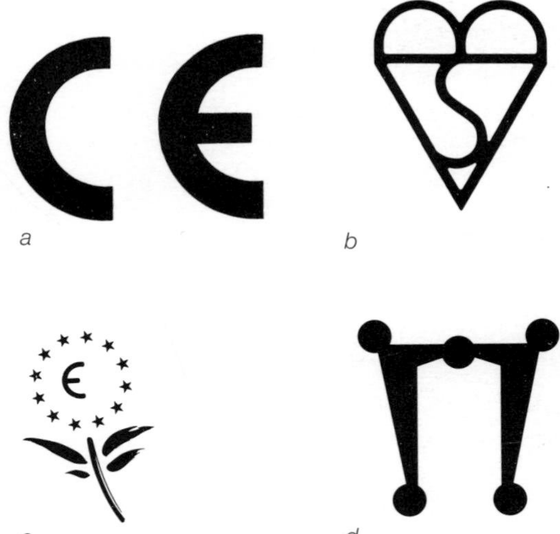

Figure **4.1**

Various types of product marks
a) CE mark
b) BSI kitemark
c) eco-label
d) CENCER mark

4.6 General design codes

There are a number of general codes to assist designers:

- vessels and tanks (BS 4994)
- rules for construction and certification of vessels less than 15 metres (Nordic boat standard)
- composite aircraft structure (JAR 25)
- limit state analysis (Eurocode 1)

Although the first three were devised for specific industrial applications, there is no reason why relevant portions should not be applied in related applications, if other codes do not exist.

Vessels and tanks

BS 4994 was compiled specifically for large vessels and tanks which could contain corrosive chemicals. It provides the methodology to undertake the design calculations and recommends specific factors of safety to cover unknown variations in design, manufacture or in-service. The minimum value of the overall design factor is eight.

Boats

The Nordic boat standard lays down design and manufacturing principles for boats up to 15 m in length. It also considers what is required for type certification (which is voluntary) and lists reinforced plastic materials approved for use under specified manufacturing conditions *(Table 4.3)*.

Table **4.3**

Approved material types (Nordic boat standard)

material type

- polyester and similar resins
- gel coat and topcoat
- thermoplastics
- sandwich adhesives
- sandwich core materials
- glass fibre products
- other fibre products

Some 600,000 boats have been certified under this scheme in the last twenty years and this experience has been carried over into the manufacture of wing blades for wind turbines.

Lloyds Register's rules for yachts and small craft incorporate a chapter on GRP construction which covers the requirements for the moulding environment and construction details as well as the structural requirements for laminates and supporting stiffeners. Materials approval requires, among other things, a minimum level of properties, a maximum scatter about the mean, an allowance for a higher level of properties in the calculations if this can be demonstrated. It is sensible to adopt such a scheme generally for all composite materials and designs.

Composite aircraft structures

The aircraft structural code for reinforced plastics is quite general. It is neither application nor process specific. It enjoys global acceptance having been devised jointly by the Aviation Authorities in both Europe and the USA. It provides an acceptable (although not the only) means of satisfying certification requirements for aircraft.

The code relies on identifying the most severe load case to be encountered in service and showing that the component or structure could continue to perform its function as the result of such a loading at the very least until an inspection could be carried out.

Structural design codes (Eurocodes)

These codes were initiated by the EC with a view to harmonizing rules within the European construction industry as part of the Construction Product Directive. The objective is to produce a set of design rules which are technically sound, unambiguous, easy to use and based on the use of limit state analysis (*Box 4.6*). The preparatory work has been under development for 10 years and the first pre-standards (ENV) have now appeared (1992).

Summary

These codes reflect the two opposing approaches to the design of load bearing components and structures. Both require the designer to characterize designs and materials. It is the variation in materials on processing and subsequent durability in service which leads to one or other approach being adopted.

The use of such codes by agreement within industry is likely to accelerate the transition from stress- to limit-state. This has already occurred for aircraft and is now under discussion for structures such as off-shore rigs and wind turbines.

Box **4.6**

Structural Eurocodes

Formation of Eurocodes

Eurocodes are being drafted on the basis of a format, which draws a distinction between principles and application rules. Adherence to principles will be mandatory, but designers will be able to use alternative methods to satisfy principles provided that the methods can be justified.

Eurocode 1 will set out a common set of principles governing the basis of design and the actions and loadings on structures. Two further codes will consider geo-technical and seismic design. These principles are interpreted in subsequent codes for various constructional materials like steel, concrete, timber and masonry.

Limit state

The ways in which a structure fails to fulfil its function need to be identified. Each is treated as a limit state in which any of the performance criteria governing the use of the structure are infringed.

The designer needs to distinguish between:

- ultimate limit states corresponding to collapse, such as loss of stability, failure of critical sections, connections or structural members by rupture, excessive deformation or other situations which jeopardize life or cause major economic loss ▶

- service limit states in which the structure can no longer fulfil its design function, such as major deflections or unacceptable vibrations

The limit state is then modelled and uncertainties associated with the variables or the model are included using safety (or partial) coefficients. The Eurocodes pre-standards recommend indicative values. Precise values may be specified by member states during the period of validity of the pre-standards (three years).

4.7 Specific design codes

In addition to the general design codes, there are a number of specific codes covering applications in which reinforced plastics have been or are being used.

Specific properties

Some codes exist by virtue of the specific properties of reinforced plastics, the material being explicitly defined – for example, translucent roofing sheeting (BS 4154), in which the key requirements are the amount of light transmitted and the mechanical strength, which determines the spacing between rafters.

Design elements

Some codes cite reinforced plastics as only one of many competing materials – for example drainage pipes (*Box 4.7*). For protective barriers (BS 6180), the emphasis is on the design elements that need to be considered. This specification does, however, lay down some very specific requirements for the use of reinforced plastics:

- structural members can only be made from reinforced thermosetting resins.
- a design factor of 1.5 is to be applied to the design loads after taking into account ageing due to long-term weathering.
- the surface spread of flame must meet Class 3 BS 476: Part 7.

Box **4.7**

Pipes, joints and fittings

GRP pipes

For drainage pipes, GRP is only one of eight possible material types which are differentiated by their ability to convey various types of liquids and effluent and withstand ground movement or slip by having some flexibility (BS 8010).

Piping materials have to be selected according to specific requirements – for example the conveyance of potable water is given in Table 4.4.

Table **4.4**

Requirements for pipes for potable water (BS 6920)

Specific requirements

- not support microbiological growth
- not give rise to unpleasant taste or odour
- produce no cloudliness
- cause no discolouration of the water
- meet both WHO and CEC guidelines for specific metals and injurious substances.

General requirements include a nominal lifetime of 50 years, but for reinforced plastics no such service data is yet available. An example of a tank after 25 years is given in *Box 1.4.*

The codes do provide methods of undertaking accelerated testing, and one such test is shown in *Figure 4.2.*

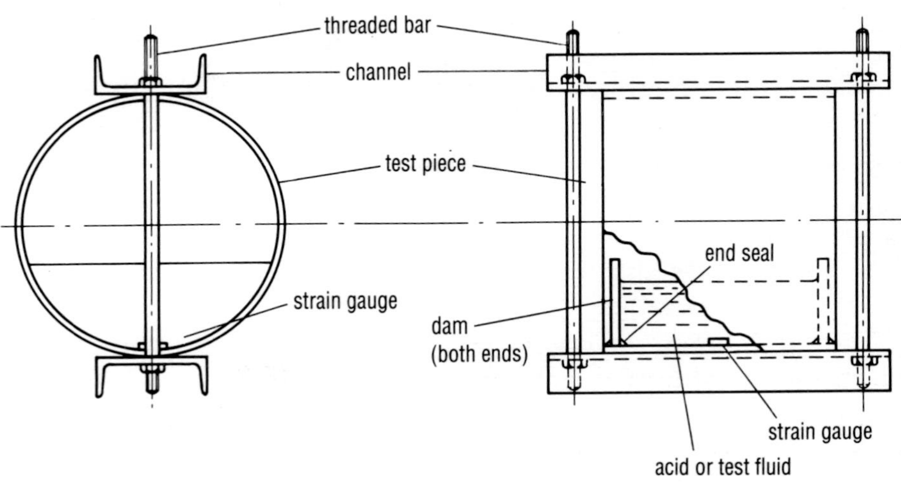

Figure **4.2**

Scheme for testing small lengths of pipe for strain corrosion. The bottom portion of the pipe acts as a reservoir for the corrosive or test fluid and the load is applied through the threaded bar (BS 5480)

The relevant codes and standards are set out in *Table 4.5.* These codes cover almost all applications where GRP might be used. They recommend among other aspects dimensions such as nominal diameter, effective length and squareness, effective service temperature, and the design and testing of pipes, fittings and joints.

Table **4.5**

Pipes, joints and fittings

plant	standard
• water supply or sewerage	BS 5480
• process plant	BS 6464
• code of practice for plants or sites	BS 7159
• sewerage: guide to new construction	BS 8005
• code of practice for pipelines (pt 2.5)	BS 8010
• code of practice for drainage	BS 8301

The flexibility of GRP pipes enables them to tolerate more ground movement than rigid pipes. However they must be handled more gently, correctly laid and trenches must be suitably tamped and backfilled to avoid damage. Steel pipes on the other hand are much less vulnerable to handling.

Glass reinforced plastic plant

Some of the major codes are listed in *Table 4.6*.

They vary as regards the type of component, the internal pressure, the working temperature and design exclusions. BS 4994 is the most general in terms of component type and static head, and ASME X permits the highest working pressure.

Table **4.6**

Vessels and tanks

plant item	standard
• vessels and tanks	BS 4994
• fibre reinforced plastic pressure vessels	ASME X
• filament wound GRP chemical resistant tanks	ASTM D3299
• offshore installations: guidance on design, construction and certification	SOLAS
• contact moulded reinforced plastic chemical resistant process equipment	NBS

Caution

The codes have in general been verified for particular types of material and methods of manufacture. If different materials or methods of manufacture are used then additional tests should be undertaken to verify the design.

4.8 Design considerations for fire

Flames induced by a fire can cause plastics to burn, thereby adding to the propagation of the fire while the heat induced or self-induced can cause the plastics to soften so that they no longer can carry the required load. Thermoplastic resins (unlike thermosetting resins) will melt if the temperature exceeds their softening point *(Section 5.3)*.

Such considerations need to be assessed during the conceptual design phase, and again in the detailed design phase where the choice of material and the design are refined *(Section 2.3)*.

Phases of development

Specific aspects that must be considered are listed in Table 4.7; Figure 4.3 shows how these are relevant to the different phases of a fire.

Table **4.7**

Flammability characteristics of fires

characteristics

- ignitability
- rate of heat release
- surface spread of flame
- smoke obscuration
- toxic and /or corrosive fume emission
- fire resistance (penetration)

Considering each phase briefly:

Ignitability concerns the time and temperature at which burning will occur under external heat and/or a small flame, while fire growth is concerned with the increasing rate of **heat release** as well as the rate of flame spread across surfaces.

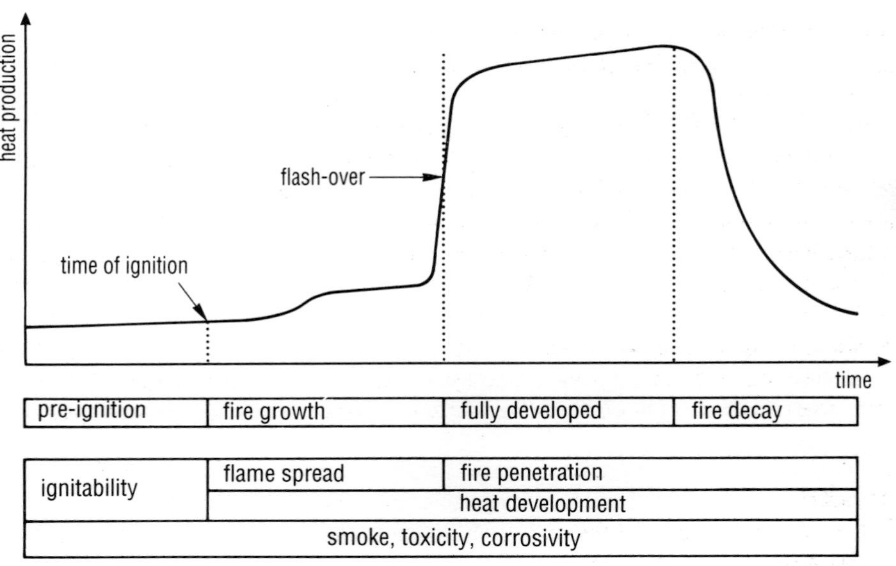

Figure **4.3**

Diagram showing the different phases in the development of a fire within an enclosed space (ISO 3814)

Surface spread of flame is the result of a fire heating the surface adjacent to it, causing volatile components to ignite progressively across the surface. The speed with which the flame front advances and the extent of travel are measured in order to define this property.

Smoke is the visible suspension in air of solid and/or liquid particles resulting from incomplete combustion. Obscuration occurs not only from the presence of smoke, but also from the effect of heat and gaseous irritants. It is regarded as the major factor in delaying effective escape from enclosed spaces.

Emission of **toxic fumes** is difficult to quantify. The lowest level of concern may simply be an irritant, while at higher levels they can affect the ability to escape from a fire, and at the highest level they can be lethal. Individual thresholds also vary considerably.

Screening tests

It is now imperative to screen materials effectively, and the various aspects that need to be considered in relation to the application are set out in *Table 4.8*. Small-scale tests are very seldom representative of full-scale tests. They are, however, useful in initial screening.

Table **4.8**

Fire test aspects of materials

considerations

- applicability of test to full scale fire performance
- size of sample and scale of test
- other sample factors like shape, surface area, thermal inertia and combination with other materials
- test conditions
- method of applying flame or radiant heat source
- presence of oxygen or ventilation conditions

Two tests are widely quoted: UL 94 and the oxygen index. UL94 assigns a flame class based on the burning behaviour of a sample of specified thickness when it is ignited with a small ignition source. The oxygen index measures the minimum oxygen concentration in an admixture of nitrogen required to sustain combustion (*Table 4.10*).

There is as yet no international agreement on codes and standards, with every country so far having its own code and ways of limiting the effects of fire. However, with the increasing understanding of the phenomenon, considerable effort is going into standardization work within CEN, ISO and IEC.

Hazard analysis and risk assessment

If fire poses a hazard then the designer should consider the level of risk that is permissible This would be different in aeroplanes and in trains, for example.

The following five-step approach to hazard analysis is advocated by BPF in its recent code of practice for the assessment of smoke obscuration hazards (*Table 4.9*).

The hazard might be temperature (flame or room), radiant heat, smoke obscuration, toxic fumes or lack of oxygen, or a combination of two or more factors. Thus the analysis has to assess the point at which the hazard becomes unacceptable. If this time is insufficient, or the danger associated with the fire too great, then either the materials or the design must be modified or changed.

Table **4.9**

Fire hazard analysis

considerations
• definition of fire scenario
• ignition phase
• fire growth phase
• test data collection
• rate of build up of hazard

4.9 Fire safety

Design for fire safety needs careful thought as it involves aspects of both design and material selection, and these cannot always be separated. As set out in *Figure 2.5*, a design strategy must be evolved to ensure an adequate margin of safety.

Consider four specific examples in which alterations to material specifications, design, and even the method of use have been imposed as the result of fire.

- a jammed contact within the switchgear of a wind turbine at Masnedø in Denmark led to a fire which burnt out the nacelle housing, which was made of glass reinforced polyester resin. The housing material was subsequently altered to a fire-retardant grade of polyester resin.
- rubbish accumulating under an escalator at one of London's Underground stations was lit by a cigarette – as a result, smoking was banned throughout the Underground system and rubbish removal was improved.
- toxic emissions from burning furnishings inside an aircraft at Manchester Airport as a

result of a fire which started externally resulted in the introduction of fire-retardant fabrics and stringent heat-release and smoke requirements for all aircraft furnishings.

- the spread of flames which occurred through the seal between tunnel sectors at London's Oxford Circus Underground station resulted in improved screening tests done on larger amounts of material and in the substitution of improved fire-resistant materials.

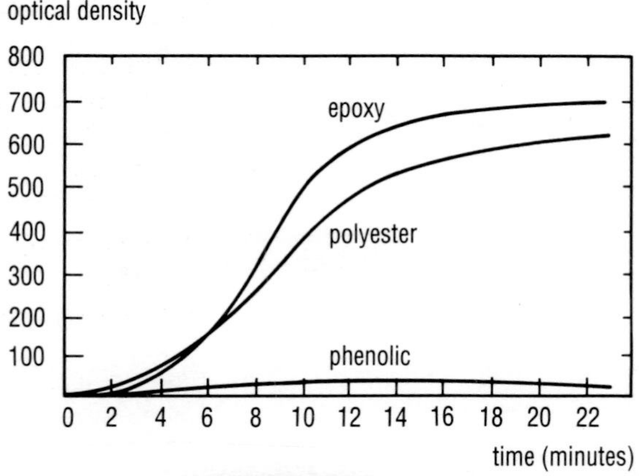

Figure **4.4**

Smoke production in the NBS smoke chamber test. The phenolic chemical structure is char forming leading to low smoke emissions while epoxy and polyester do not form a char (Forsdyke)

Fire resistance

The three prime factors to obtain fire resistance are stability, structural integrity, and heat conductivity.

Stability must be maintained. Reinforced thermosetting resins will not carry any appreciable load near their heat distortion temperature, while thermoplastic resins will melt.

Integrity needs to be preserved while a fire is being fought and an evacuation is in progress. The latter will depend upon the design of stairwells and walkways, the materials used to construct them, and the time required to evacuate personnel or passengers.

As reinforced plastics conduct heat poorly, they will be cooler to the touch immediately away from the heat source (unlike metals, which are excellent conductors). GRP piping is being used on tankers and being considered for offshore use. This property needs to be balanced against the need to meet other requirements such as surface flammability, toxicity and smoke obscuration.

Fire retardancy

This can be achieved in a number of different ways and these need to be considered in relation to the relevant fire standards *(Box 4.8)*:

- some resins are inherently less ignitable and less smoky than others, as the example in *Figure 4.4* shows for three thermosetting resins.
- additives can be mixed with the more flammable resins, but care must be taken that the additives do not give off toxic fumes or increase smoke production which would be unacceptable in a confined space or escape-way where persons could be present.
- intumescent paints are surface coatings based on polyester or epoxy resins which are designed specifically as a protective coating against direct flame. They are formulated so that when a flame is applied to a cured surface, a carbonaceous foam is formed and the inert gases produced insulate the underlying surface against the flame *(Figure 4.5)*. Their use is not restricted to reinforced plastics, and they are widely used to protect structural steelwork.

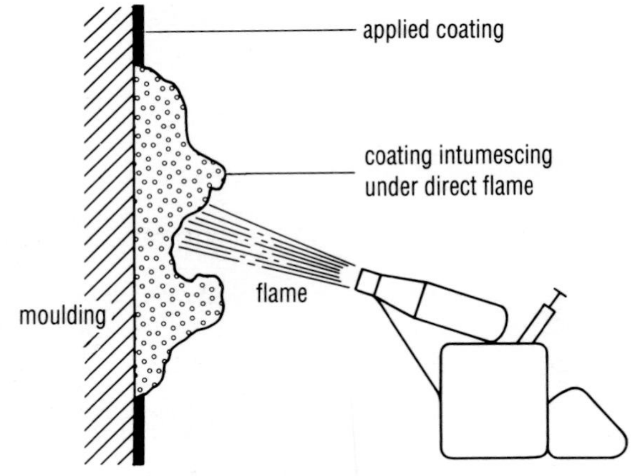

Figure **4.5**

Action of a flame on an intumescent layer creating an insulating foam to protect the underlying surface (Scott Bader)

Box **4.8**

Tests

Reaction to fire tests

There are many standards and tests currently in use in various countries. The major ones are listed in *Table 4.10* together with the emerging European and international standards. Various classes are defined for each test, which depend on the rate of advance (as a percentage of the flame front) or an amount of fire effluent emitted.

These are often prescribed by regulation or within the design brief. If fire is perceived to be a hazard, testing may currently (1992) have to be carried out by approved test houses in each country of use.

Table **4.10**

Fire tests

property	standard	country (comment)
non combustibility	ISO 1182	International
	BS 476 pt 4	British
ignitability	ISO 5657	International
	ONORM B3800 pt 1	Austrian
	NT fire 002	Nordic countries
flame spread	ISO DIS 5658	International
	BS 476 pt 7	British
	NEN 3883	Netherlands
	NT fire 004	Nordic countries
	DIN 4102 pt 1	West Germany (small burner)
	DIN 4102 pt 15	(large burner 'brandschact')
	pt 16	
	NFP 92501	France (epiradiateur)
	UL 94	USA: Underwriters' laboratory
heat release	ISO 5660	cone calorimeter
	BS 476 pt 6	British; classes set by UK building regulations
smoke release	ISO/TR 5924	dual chamber
	ASTM E 1354	USA; cone calorimeter
	ASTM E 662	NBS smoke chamber
	BS 6401	NBS smoke chamber
	SIA 183/2 pt 2	XP2 smoke chamber
oxygen index	ASTM D 2863	USA
	BS 2782 pt 1	method 141; British
	NT FIR 013	Nordic countries
toxicity index	DIN 53436	Germany

Notes:

- *Toxicity index is a multiple of gas concentration and a toxicity factor to produce an index which weights most toxic species generated.*

Table **4.11**

Specific fire tests

product	standard	comment
building regulations	NT fire 025	room test
	BS 476 pt 3	roofs (external exposure)
	ASTM E 162	small scale test
	ASTM E 84	large scale test
	ISO 9705	large scale test
railway rolling stock	BS 6853	3m cube smoke test; flammability temperature index motor
vehicles	FMVSS 302	flame spread
shipping	IMO res A653(16)	heat release, flame spread
aerospace	ATS 1000.001	flammability, smoke generation, toxic gas emission
	FAR 25.853	heat release, smoke
	JAR 25.853	

Notes:
- *FMVSS 302 is a USA test (Federal Motor Vehicle Safety Standard), but is now in almost universal use for motor vehicle interiors.*
- *ATS is an Airbus Industrie test, also used by other manufacturers.*
- *FAR is a Federal Aviation Regulation published by the American Federal Aviation Authority; JAR is the British equivalent.*

Evacuation time

Time is crucial if people have to be evacuated. It is the rate at which heat builds up and smoke or toxic fumes are produced that is important. Anything which can reduce these rates should be considered – for example fire walls to prevent the spread of flames.

The onus rests with the designer to consider the worst possible circumstances that could arise and design accordingly. If fire is possible then decisions about structural and fire retardancy need to be considered before starting the detailed design. Such decisions could affect the choice of materials, mechanical properties and possible processing routes.

Material selection

Fire-retardant grades of **polyester** are available of which the more common formulations are ex stock. They are produced by using:

- a suitable inert filler
- a halogen agent to the molecular backbone or an additive
- a water releasing agent

Phenolic resins have inherently good fire properties because of char formation. Fire retardant grades of **epoxy** are generally only available for specialized applications such as printed circuit boards. Some **thermoplastics** are inherently fire retardant, including polyphenylene sulphide, polycarbonate, PEEK.

Note that the higher the maximum operating temperature of a resin *(Section 5.3)*, the higher the margin of safety.

4.10 **Reference information**

4.1 **Single market**

Treaty on European Union, CEC, Luxembourg, 1992 – text of Maastricht treaty signed 7.2.92.

Economics of the European Community, 3rd ed – El Agraa (ed), Philip Allan, London, 1990 – history of the various institutions, background to the single market.

Single Market, the Facts, DTI, London, 1992 – six-monthly update on all aspects of the single market.

Guide to the Sources of Advice on the Single Market, DTI, London, 1992; other information via DTI Single Market 'hot line' (44) 081 200 1992.

Completing the Single Market: the removal of technical barriers to trade within the EEC, CEC, DG III/B/4, 1990.

'Role of European standardization in the European economy' EC Resolution 92/173, *Official Journal of the European Community* 9.7.92.

4.2 **Liability**

The approximation of the laws, regulations and administrative provisions of member states concerning liability for defective products, EC directive L210/29, *Official Journal of the European Community* 7.8.85.

Guide to Consumer Protection Act 1987 – product liability and safety provisions, DTI Consumer Safety Unit, 1987 – brief description.

Product Liability, IMechE, London, 1989 – brief guide providing advice to both individual designers and firms.

Product Liability, C J Wright, Blackstone, 1989 – the law and its implications for risk management.

Safer by Design, H Abbott, Design Council, 1987 – management of product design risks under strict liability.

4.3 **Safety**

Health and Safety at Work Act, Health and Safety Executive, London, 1990 – brief outline of the act.

Essentials of Health and Safety at Work, HSE, 1990 – practical guide; reference section on guidance notes for specific applications, processes and products.

Health and Safety, I Fife and E Machin, Butterworth, London, 1990.

'General product safety', EC directive 92/59, *Official Journal of the European Community 228/24 BSI News*, 10-11, September 1992, for an explanatory comment.

4.4 **Standards and codes**

CEN documents are available as transposed national documents. Annual CEN publications include a standard catalogue and list of draft standards.

Bulletin of the European Standards Organisations and *BSI News* are published monthly – providing useful updates.

New Approach Directives, DTI, London – explanatory booklet on each, updated as necessary.

'Single market programme, public procurement', *BSI News*, July 1992 – status of relevant EC directives and incorporation in UK law.

4.5 **Quality systems**

Quality Management and Quality Assurance Standards: Guidelines for Selection, ISO 9000, 1987; *ISO 9001* to *ISO 9004* cover specific aspects; the equivalent BS standard is the various parts of *BS 5750*.

Environmental Management Systems, BS 7750, 1992.

'Community Eco-Audit Scheme for Companies', Draft EC Regulation 76/2, *Official Journal of the European Community*, 27.3.92.

'Indication by Labelling and Standard Pproduct Information of the Consumption of Energy and Other Resources of Household Appliances', EC Directive 92/75, *Official Journal of the European Community*, 13.10.92.

CE mark: refer to discussion in 'Construction Product Directive: Information for Manufacturers and Suppliers', *Euronews Construction*, DoE, London, September 1991; consult also national or European trade association, local trading standards authority or DTI/DoE contact.

Product marks – consult relevant national body like BSI, Milton Keynes, UK +44 0908 220 908.

Eco-labelling (*Section 2.11, loc cit*).

4.6 **General design codes**

Nordic Boat Standard: Rules for Construction and Certification of Vessels Less Than 15 metres in Length, Det Norske Veritas, Høvik, 1992.

Composite Aircraft Structure, (Acceptable Means of Compliance), JAR 25.603, Civil Aviation Authority, Gatwick, 1986.

Design and Construction of Vessels and Tanks in Reinforced Plastics, BS 4994, 1987 – under consideration and review as a CEN standard (1992).

'The New Generation of Structural Eurocodes', *Euronews Construction*, DoE, 1992 – current status of Eurocodes and notes on the way these will be implemented.

Design of Concrete Structures, ENV 1992-1-1 (Eurocode 2); *Design of steel structures, ENV 1993-1-1* (Eurocode 3) – part 1 general rules and rule for building – now published as drafts for development by BSI, 1992.

The Basis of Design and Actions on Structures, ENV 1991 – basis of limit state analysis is still in drafting stage.

4.7 **Specific design codes**

Lists of BS and ISO standards are published annually and are available from BSI, Milton Keynes.

ASTM standards are listed in *Vol 8: Plastics* – ASTM, Philadelphia, published annually.

'Glass Reinforced Plastics Plant', F E Lawrence in *Reinforced Plastics,* 1983 p44-58.

4.8 **Design considerations for fire**

Code of Practice: Assessment of smoke obscuration hazards caused by plastic products in fires, British Plastics Federation, London, 1989.

Fire tests for buildings, products and materials, A D Makower, BSI, Milton Keynes, 1989 – summary of some 300 British standards dealing with fire test methods and specifications.

Fire reaction tests in the EEC, G Blachre et al, CEC DGIII/B/5 No 3197, 1989 – can a material be accepted in a member state on the basis of the results performed in another member state?

4.9 **Fire safety**

'Phenolic FRP today', K L Forsdyke in *Proc. BPF Conf,* paper 4, BPF, London, 1989.

Polyester Handbook, Chapter 7, Scott Bader, Woolaston, 1986.

Fire Toxicity of Plastics, RAPRA, Shrewsbury, 1989 – symposium on current status of toxicity.

Fire Performance of Materials in Mass Transport, RAPRA, Shrewsbury, 1988 – symposium on current areas of concern in transport.

International Plastics Flammability Handbook, J Troitsch, Hanser, Munchen, 1986 – principles, regulations, testing and approval.

Design Data for Reinforced Plastics, N Hancox and R M Mayer, Chapman & Hall, London, 1993 – Chapter 8 describes some of the key test methods.

Fire Safety in Buildings, H L Malhotra, Building Research Establishment, Garston, 1986.

Fire Test Procedures, International Maritime Organization (IMO), London, 1984 – test methods, procedures and hazards in accordance with the Safety Of Life At Sea convention.

Tests for Measuring Reaction of Fire to Building Materials, ISO TR 3814, ISO, Geneva, 1989 – reviews risk assessment and introduces concept of a tool kit of tests.

Toxicity Testing of Fire Effluents, ISO TR 9122, ISO, Geneva, 1989 – reviews current state of the art and how such tests may contribute towards assessing the hazard.

Guide for the Assessment of Toxic Hazards in Fire in Buildings and Transport, DD 180, BSI, Milton Keynes, 1989 – provides guidelines for assessing toxic hazard and gives worked example of a warehouse.

Data sources for combustibility of materials

Some limited information is given on databases such as those of RAPRA Technology *(Section 5.6)*, which can serve as an initial guide.

If fire is a possible hazard, then the material supplier should be consulted about the precise formulation of the resin or compound and whether it would meet specific regulations. In general, testing may have to be carried out by approved test houses in each country of use.

Independent advice can be sought from test houses in the UK such as RAPRA Technology (loc cit) or SGS Yardsley, Trowers Way, Redhill, Surrey, RH1 2JN, UK +44 0737 765 070, and there are independent consultants who can advise on design strategies.

Bibliography source for standards and codes

PERINORM contains up-to-date and detailed bibliographic entries on all the current, full and draft standards and specifications for the UK, France and Germany, plus all European and International standards produced by ISO, IEC and CEN/CENELEC, updated monthly. In addition there are records for the technical rules and regulations that are applicable in France and Germany. A search can be carried out in English, French or German, and is based on a CD disc for use with a personal computer. A similar CD-ROM based system is Standards Infodisk, which currently (1992) covers standards issued by 115 organizations world-wide.

Within the UK, consult a reference library such as that of the Institution of Mechanical Engineers, 1 Birdcage Walk, London SW1H 9JJ +44 071-222 7899 for assistance with codes and safety.

For aspects concerning health and safety, consult Health and Safety Executive, Belgrave House, 1 Greyfriars, Northampton, UK +44 0604 21 233.

Certification societies such as Lloyds Register, 71 Fenchurch Street, London EC3M 4BS +44 071 709 9166, and Det Norske Veritas, 112 Station Road, Sidcup, Kent DA15 7BU +44 081 309 7477 are able to provide advice concerning structural design.

Structural design service

Lloyds Register offer such a service for builders and designers of yachts and small craft. The basic service involves the computing of either the single skin or equivalent sandwich laminate requirements in conjunction with internal stiffener arrangements. Either calculated mechanical properties or those determined from tests on representative production laminates are used.

Standards

Both British Standards and other material and international standards can be purchased from BSI, Linford Wood, Milton Keynes MK14 6LE +44 0908 220 022.

European standards being issued by national bodies will normally carry the prefix BS for the English version or DIN for the German version (eg BS EN 123).

International standards adopted by CEN are issued in the general series as 20 000 plus the ISO number (eg ISO 9000 has been adopted by CEN as EN 29 000).

Some ISO Standards appear in the form of equivalent national standards, for example ISO 178 is equivalent to BS 2782 part 3 method 335A. Numbers in the UK BS and the ISO are unrelated and so the BSI catalogue must be consulted.

Materials and properties

chapter 5

summary

This chapter provides a brief overview of the types and properties of individual constituents of reinforced plastic materials. These are not necessarily the starting points in a design, so a similar survey is made of semi-processed materials such as fabrics, moulding compounds and prepregs.

The chapter also looks at the behaviour of a typical fibre reinforced plastic to highlight the differences between such materials, plastics and metals.

Introduction

Fibre reinforced plastic (FRP) materials consist of stiff, strong fibres bonded together by a plastic or resin matrix. The fibres enhance the low stiffness and strength of the resin. The resin's main purpose is to transmit loads into the stiff, brittle fibres and to protect them from damage.

Selecting compatible fibre/resin systems (with appropriate fabrication methods for positioning fibres) can result in engineering materials that combine the high stiffness and strength of the fibres with great toughness, since the reinforcing fibres inhibit crack propagation in the resin matrix.

Not all fibre/resin combinations lead to FRP materials with these enhanced properties, nor ones which may be readily processed into composite components. This chapter therefore provides a guide to the fibres, resins and composites currently in use.

FRP materials may be classified by fibre type and resin type. However, a more practical classification is based on fabrication technology. This gives three main classes of materials:

(a) separate fibres and resins for the user to process into a composite by fabrication.
(b) compounds with the correct mixture of fibres, resin and other additives ready for press or injection moulding into a component.
(c) finished items ready to assemble into a component, such as pultruded sections (*Section 2.6*), or filament-wound pipe.

For type (a) materials, the user has complete freedom to choose fibre reinforcements, proportions and fibre orientations to tailor the properties of the resulting composite. The properties of type (b) materials will depend on moulding conditions, but are otherwise fixed by the compound supplier, so the user has less involvement with fibre/resin chemistry. Type (c) materials have already been discussed in *Section 2.6*.

This chapter briefly describes the main constituents of fibre reinforced plastics, and reviews the main properties of the resulting composites.

5.1 Fibres

The three main types of fibre used for reinforcing plastics, in order of increasing strength and cost, are glass, aramid and carbon. They are described in *Boxes 5.1* and *5.2*, together with some current applications.

Properties

Typical physical properties for these fibres are listed in *Table 5.1* and their comparative stiffness and strength properties are shown in *Figure 5.1*.

Two derivatives are shown for each type of fibre to indicate the range of properties available. One striking point is the small (and negative) coefficient of thermal expansion for aramid and carbon fibres compared with those for glass fibres or metals. This can be extremely useful if a mechanism must have good thermal stability.

5.1 **Fibres**

The favourable ratio of stiffness and strength to density compared with metals has already been discussed *(Section 2.8)*, and has been used to advantage in many applications *(Section 1.2)*.

Table **5.1**
Typical physical properties of fibres

	Density (Mg/m^3)	Thermal expansion (x10^{-6}/K)	Thermal conductivity (W/m.K)	Electrical volume resistivity (Ohm.m)
E glass	2.5	5	1.0	4×10^{14}
R glass	2.5	4	1.0	
aramid				
Kevlar 29	1.44	-2	3.0	
Kevlar 49	1.45	-2	3.0	5×10^{13}
HM carbon	1.9	-1.3	105	8.7×10^{-6}
HS carbon	1.8	-0.3	24	14.0x10-6

Note: for other physical properties refer to Figure 5.1. Kevlar 29 is also designated intermediate modulus or high strain. Kevlar 49 is also designated high modulus (HM). Carbon fibres may be designated HM or high strength (HS).

Box **5.1**
Glass fibres

Form and usage
Glass is the principal fibre reinforcement for plastics, supplying about 98% of the fibre reinforcement market. It is used as continuous fibre reinforcement primarily with polyester and epoxy resins, in chopped form as reinforcement for thermoplastic injection moulding materials such as nylon and polypropylene, and in various types of fabric *(Section 5.2)*.

Glass fibres are relatively cheap, have a high strength-to-weight ratio, and good thermal and electrical insulation. They are easily handled and processed into composite components using a wide range of fabrication techniques. As a result, glass fibre reinforced plastics (GRP) are widely used in all the main industrial sectors, in particular for boat building, chemical plant pipework, food storage vessels, building panels, automotive components and sports goods.

Types of glass fibre
Various glass compositions are available in fibre form and these are listed in *Table 5.2*

Table **5.2**
Different types of glass fibre

fibre type	principle use
E	standard reinforcement, low alkali content (<1%)
A	high alkali content (10 – 15%), inferior properties to E and not widely used
C	improved corrosion resistance to E, normally used in form of surface tissue
E-CR	boron-free, good acid corrosion resistance, other properties similar to E
D	high silica and boron content, dielectric applications, radio frequency transparent
R, S	better mechanical properties than E but higher cost, so specialized applications
AR	alkali-resistant, for reinforcement of cement

Box **5.2**

High performance fibres

Aramid and carbon fibres are the principal high performance fibres. They are more expensive than glass fibre *(Section 2.9)* and each has less than 1% of the reinforcement market by volume, though their share is expected to grow.

Aramid

Aramid fibres generally have higher tensile stiffness and strength than glass fibres *(Figure 5.1)* and a lower density. They are used mainly as continuous fibre reinforcement in epoxy resins for specialist applications where higher stiffness and lower weight than GRP materials are needed, for example in sports goods, power boats and aerospace components.

Aramid composites are tough, with good thermal stability and impact energy absorption properties, and are thus used in specialist applications such as body armour, helmets and lightweight armour plating. They are not considered suitable for applications involving high compression and flexural stresses as they have relatively low compression strength.

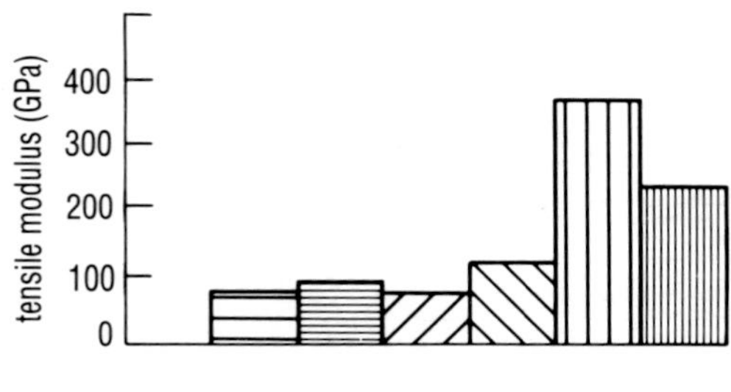

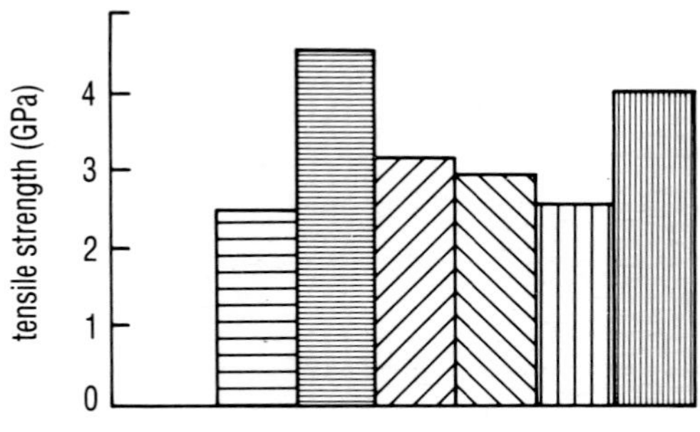

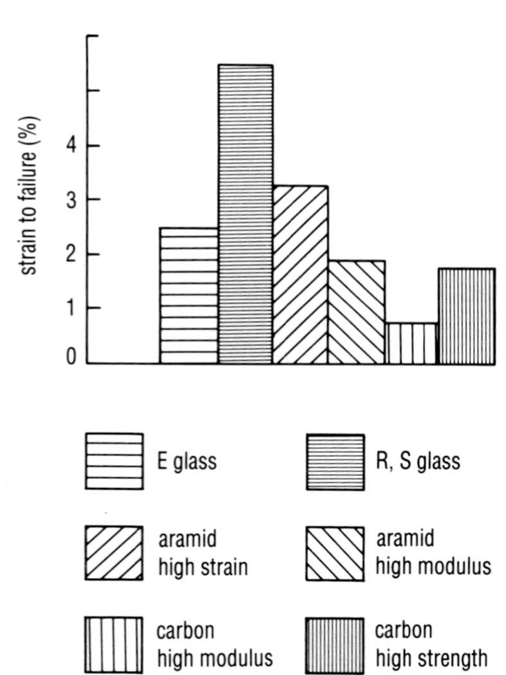

Figure **5.1**

Tensile modulus, tensile strength and strain to failure for some reinforcing fibres (after Quinn)

Carbon
Carbon fibres have the highest combination of stiffness and strength properties of the three principal fibre types *(Figure 5.1)*. They are used mainly as continuous fibre reinforcement for epoxy resins, and more recently in thermoplastic resins such as PEEK. Some chopped fibres are also used in thermoplastics such as polyamide 66 for injection moulding.

Carbon fibre reinforced plastics (CRP) are used primarily in applications where their high stiffness to weight ratios with good creep and fatigue resistance are required, and where costs are not critical or weight saving is essential.

The main market is in aerospace applications such as space and satellite structures, military aircraft wing and fuselage panels, and internal fittings such as seats and floors in commercial aircraft. The second largest market, which uses about 30% of carbon fibres, is in high performance structures such as power boats and racing cars, and for sports goods such as tennis racquets *(Section 8.4)*, skis and golf clubs *(Section 8.2)*.

Other fibres
Further fibre types available commercially and used in small quantities for specialized applications include polyethylene, ceramic and boron.

Polyethylene (PE) fibres are produced from highly oriented PE molecules being extruded through a die. They have low density (0.95 Mg/m^3), high tensile modulus (115 GPa) and high tensile strength (3 GPa). They are being used in certain applications to replace aramid fibres, but they have a low melting point and are not suitable for applications above 100°C.

Ceramic fibres such as alumina and alumina silica combinations are used in high temperature applications up to 1400°C, mainly as reinforcement for metals such as aluminium alloys, magnesium and titanium. Typical properties include density 3.3 – 4.0 Mg/m^3, tensile modulus 300 – 380 GPa and tensile strength 1.4 – 2.0 GPa.

Silicon carbide fibres also retain their modulus and strength properties to high temperatures (1,000°C). Typical properties are density 3.3 Mg/m^3, tensile modulus 190 GPa and tensile strength 1.5 GPa.

Boron fibres were one of the earliest reinforcement fibres and have been mainly superseded by carbon fibres, but are still used in certain aerospace and sports goods applications. Properties are comparable to carbon, but with higher compression strength. Typical values would include density 2.6 Mg/m^3, tensile modulus 480 GPa and tensile strength 3.4 GPa. Both boron and silicon carbide fibres have a larger diameter than glass or carbon fibres and so cannot be woven using textile technology.

Selection

Fibres are typically from 6 to 20 μm in diameter and are aggregated into bundles for easier handling. These are specified by the characteristics set out in *Table 5.3*.

Table 5.3
Terms which characterize fibre bundles

term	description
twist	bundles may be either twisted (yarns) or untwisted (tows or strands)
size	protective coating which aids handling
finish	coating used on filament bundles to aid coupling to resin and to preserve integrity of bundle
denier	weight in grams of 9000 m
tex	weight in grams of 1000 m

Size and finish are proprietary compounds, and are usually based on silane for glass and epoxy for carbon. The advice of the manufacturer should be sought on the compatibility (and hence the wet-out) between the fibre and the resin. Finish is usually applied only to yarns and tows intended for weaving. The finer the fibre, the bundle and the fabric, the higher the level of properties – and the higher the cost. The designer has to assess the extent to which this can be justified.

The fibre **length** affects both the method of processing components *(Section 6.1)* and their properties *(Section 5.5)*. Short and medium fibres (up to say 25 mm) are made into compounds *(Section 5.4)*, while longer fibres can be fabricated into mats or fabrics, or used directly in filament-wound structures *(Section 6.6)*.

5.2 Mats and fabrics

Mats and fabrics facilitate the handling, placing and orientation of fibre reinforcement in many fabrication processes such as contact moulding and compression moulding *(Sections 6.2 and 6.3)*. A mat usually consists of chopped or continuous fibre tows randomly orientated and held together by a binder or sometimes by needling. Their use has increased in recent years as they provide the designer with great flexibility in locating the reinforcement in new but repeatable ways.

Construction

Fabrics and mats can be made in a number of ways, each of which produces a particular type of construction with its own properties, pattern potential and economics of production *(Figure 5.2)*:

woven fabrics are the most common form, and a number of standard styles are available. The yarns are crimped where they interlace *(Figure 5.3)*. Abrasion during weaving is quite high, so textile finishes are often added by the yarn manufacturer to protect the fibres during processing. These are then burnt off and the fabric is resized by the weaver. Some different types of woven fabrics are described in *Box 5.3*.

braided fabrics are similar to woven ones but incorporate a plain pattern. They are relatively slow and costly to make, and are only available as narrow tubes and flat fabrics used for specialist applications like tubes, where fibres can be orientated at optimum angles for withstanding torque *(Figure 8.3)*.

stitch bonded fabrics have their tows held together by chain stitching with a fine thread, usually polyester.

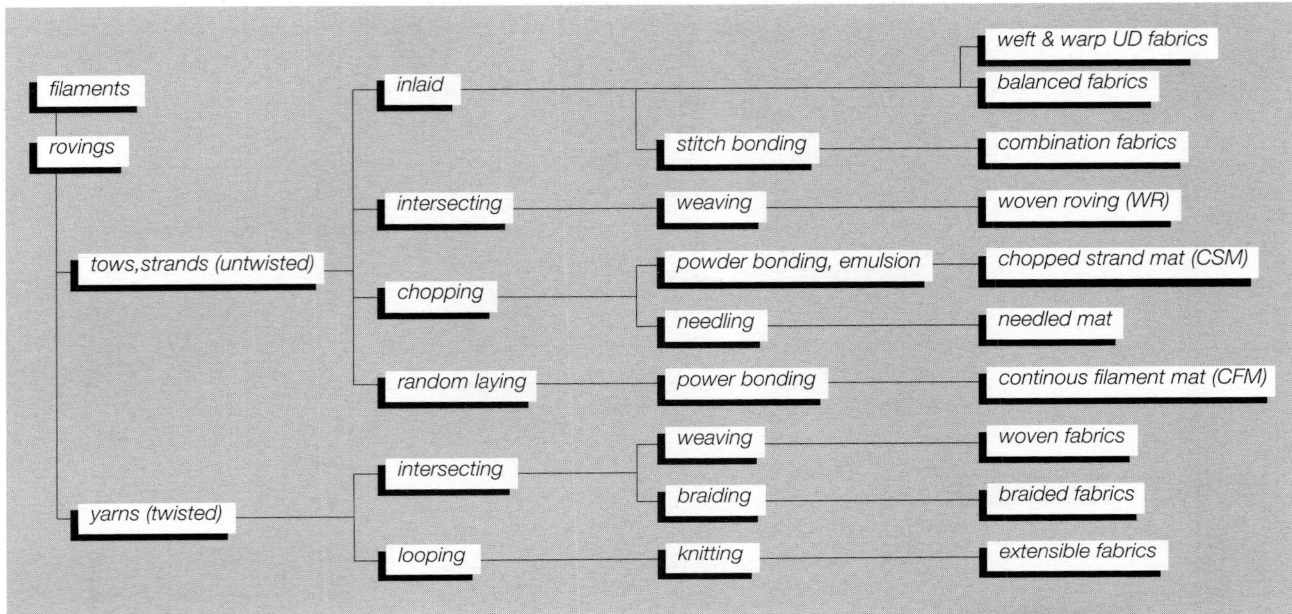

Figure **5.2**

Classification of fabrics and mats by process and type of roving; the latter may be either untwisted (strands or tows) or twisted (yarns). Powder bonding, needling and stitch bonding are three processes which can be used to form combination fabrics or mats. Other options are possible, but they are not common, mainly because of cost. (diagram based on Scott Bader)

needled fabrics are made using felting needles to interlock the tows mechanically to produce a mat.

powder bonded fabrics use a thermoplastic binder, which is soluble in resin, to bind the fibres together.

inlaid fabrics are less common than woven ones, with few standard styles. The tows are not crimped, but inlaid and secured by interlaced loops of thread (see *Figure 5.5*). Abrasion is lower than in weaving.

knitted fabrics permit considerable flexibility in yarn placement; for example, it is possible to knit-in holes or make very open structures like nets which are still stable or highly extensible fabrics. Three-dimensional fabrics are also possible. Warp knitted fabrics are almost always used (*Box 5.4*).

crocheted fabrics are made by a process which is similar to knitting but is gentler on the yarns and slower. It is best for tapes, but complex shapes are possible.

Box **5.3**

Mats and woven fabrics

Mats

Mats are one of the main glass reinforcement types. Two forms of random mat exist:

- chopped strand mat, consisting of chopped tows, typically 25 – 30 mm in length
- continuous filament mat, comprising a single length of continuous fine yarns, sometimes referred to as swirl mat.

In both types, fibre bundles are distributed in a random pattern in the plane of the mat and held together by a small amount of resin binder, which may be in either emulsion or powder form. This gives isotropic properties within the laminate plane. ▶

Woven fabrics

Plain weave glass fabrics are made from filament tows, usually with an equal number of ends and picks and are fairly thick *(Figure 5.3)*. If each layer is aligned in same direction, then strength and stiffness is higher in the fibre directions, giving the laminate orthotropic properties *(Figure 5.8)*. Good drape properties can be achieved through the use of tows rather than yarns.

Woven fabrics are widely available for all types of yarns and in various constructions with different amounts of crimp. Twill *(Figure 5.4)* and satin have better drape than plain weave but are more expensive and not so easy to handle.

Varying amounts of yarns may be contained in the warp and weft directions, ranging from balanced fabrics with equal numbers of picks and ends to unidirectional fabric in which the the warp yarns are held together by a light, open-spaced weft yarn.

Multilayer woven fabrics are used for specialised applications such as foam core sandwich laminates.

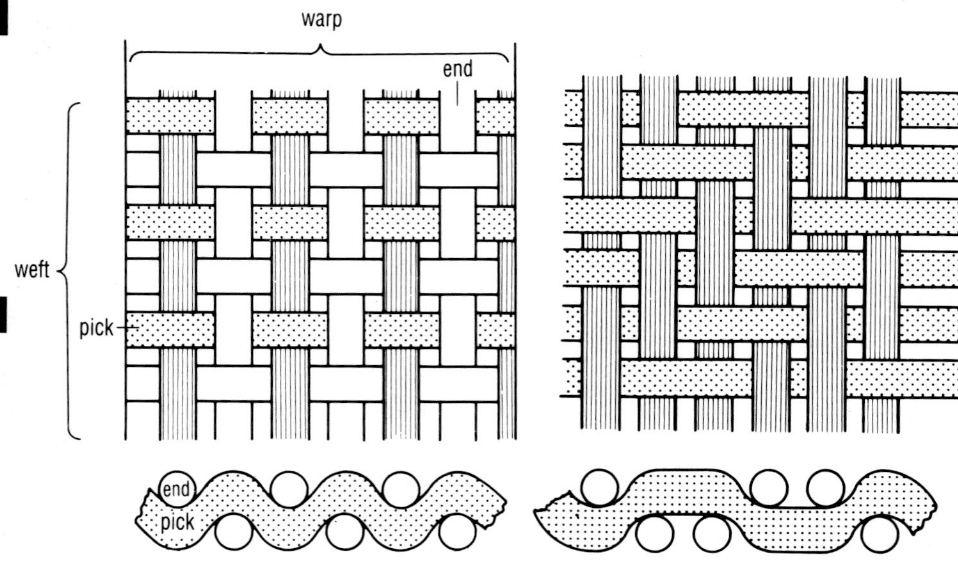

Figure **5.3**

A plain weave fabric showing how yarns of one series (weft) pass over and under the yarns of another series (warp); ends refers to the number of warp yarns and picks to the number of weft yarns

Figure **5.4**

Woven fabric of the twill type; warp and weft yarns cross in a regular sequence of 2 x 2 in this example

Box **5.4**

Inlaid, knitted and other fabrics

Hybrid fabrics are generally woven fabrics in which either each layer is a different type of fibre, or the warp and weft fibres within a layer are different, for example a carbon/glass hybrid fabric.

Inlaid fabrics

Warp (or Raschel) fabrics are balanced bi-axial **inlaid** fabrics *(Figure 5.5)*. They are equivalent to a plain weave except that the tows are not crimped but are held in place by a looped fine thread.

Variants have tows running in only the warp direction (unidirectional) or weft direction (transverse filament, for an application see *Box 1.2)*, and even multi-axial fabrics with balanced diagonal tows in addition to the warp and weft directions *(Figure 5.6)*.

Knitted fabrics
Made by looping the reinforcement yarns. This leads to fabrics with specialized properties of drape and direction of reinforcement. They tend to be designed for specific requirements and are therefore expensive compared with other fabric types.

Braided fabrics
Consist of interwoven yarns in a diagonal formation in which the yarns are oriented at equal angles $+\theta°$ and $-\theta°$ to the warp direction. They are usually produced in tubular form with the yarns inclined to the tube axis. Braiding produces a three-dimensional fabric which may be used to reinforce composite tubes or circular structures, for example propeller blades or fishing rods. Components produced from such materials have good shear resistance and torsional rigidity.

Combination fabrics
Usually consist of a layer of chopped fibre added to an underlying fabric by powder bonding, stitching or needling *(Figure 5.2)*.

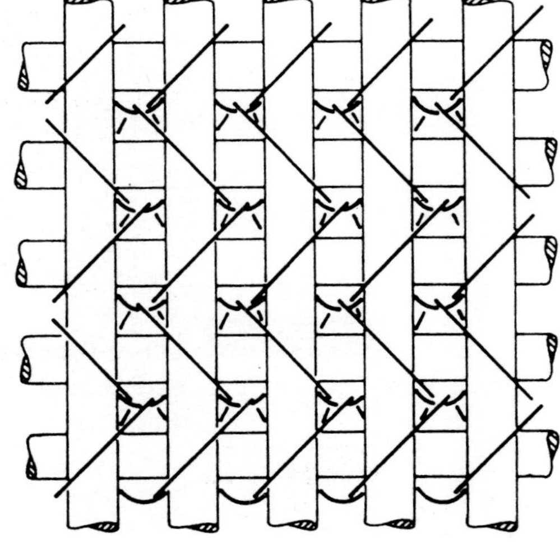

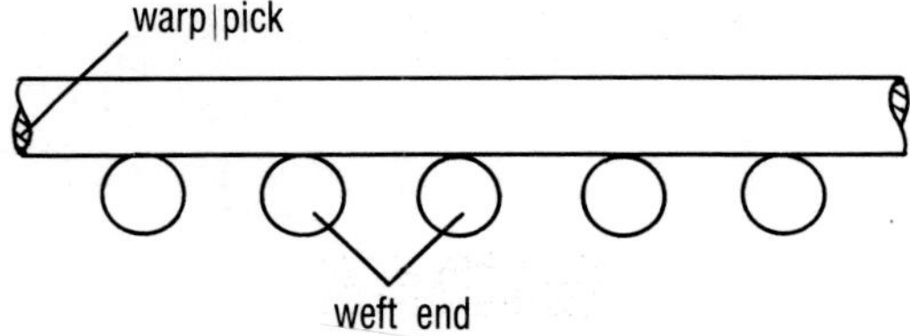

Figure **5.5**

Balanced warp/weft inlay fabric showing zero crimp of tows as these are secured by looped threads (Raz)

Selection
Some of the characteristics by which fabrics can be selected can be quantified. Others can only be understood in a general way and the materials need to be handled and tried on specific shapes *(Table 5.4)*.

Yarn orientation depends on method of construction, type and directions of fibres, and affects drape and handling.

Drapability is the ability to conform to the shape of a component. Conformity to compound curved surfaces is now possible. If the required drape cannot be obtained then either the component must be modified or the fabric must be tailored.

Handling depends upon fabric mass and style; for example, heavier fabrics and plain weaves are easier to handle than light fabrics and twill weaves without disturbing the even distribution of the yarns.

Composition may be either a mixture of short and long yarns, usually of the same type, or two different types of yarns (hybrid fabrics) to provide a better balance of properties.

Table **5.4**

Characteristics of fabrics

fabric characterisation

- yarn orientation
- drapability
- handling
- composition
- compressibility
- stackability
- price
- wetability

Compressibility is the extent to which a fabric can be compressed from its original thickness. It is useful for coping with a change of thickness.

Stackability is the degree to which layers can be combined to produce a specific orientation or fibre/matrix ratio.

Price depends on many factors, including method of construction, fibre type and availability, and fineness of yarn.

Wetability is the ability of the fabric to absorb a resin to give a good fibre/resin bond. It depends on the size and finish of the constituent fibres and the composition and mass of the fabric. Loosely woven, light fabrics wet out more easily than heavy, tightly knitted ones.

In addition to these characteristics, woven fabrics are generally specified by the parameters given in *Table 5.5*.

Table **5.5**

Typical woven fabric specification

fabric specification

- basic construction
- yarn composition(s)
- yarn type(s)
- (warp) ends (per cm)
- (weft) picks (per cm)
- width
- mass (g/m2)

Further information will be required for other types of fabrics and mats, as well as ones that are needled or powder bonded. Some fabrics are available in narrow widths as tapes for easier handling or local reinforcement. The selection process begins with choosing the

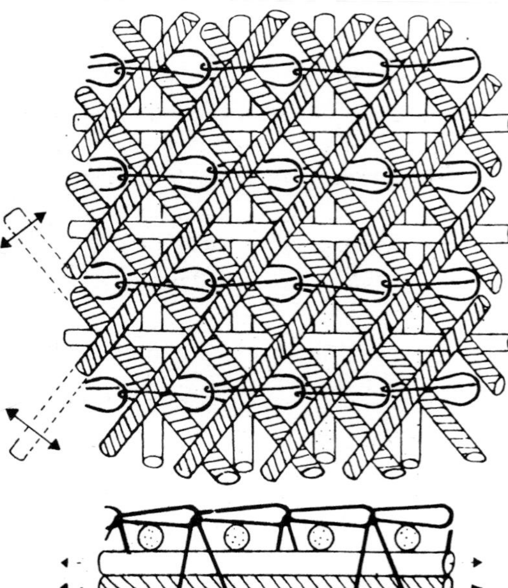

Figure **5.6**

A multi-axial inlaid fabric in which the four sets of tows are held by looped threads (Raz)

type of fibre(s) and yarn(s) to provide the desired range of properties which will be compatible with the resin and the fabrication process to be used. Simpler fabrics and mats, including some hybrid fabrics, may be chosen from manufacturers' catalogues. Special fabrics and complex shapes can only be justified if they are to be used in large volumes or there is no other way of making a component.

5.3 **Resins**

The resin is the polymeric matrix into which the fibre reinforcement is embedded. There are two principal types: thermosets and thermoplastics. The choice depends mainly on the most suitable processing route, as both types have broadly similar mechanical properties. Some of these factors are listed in *Box 5.5*.

Box **5.5**

Thermosetting and thermoplastic resins

Resin types

Thermosetting resins such as polyester and epoxy resins are usually supplied as viscous syrups which set to a hard solid when activated by a catalyst, and remain solid on further heating. Reinforced thermosets are the basis of traditional, established fibre reinforced plastic materials.

Thermoplastic resins such as polypropylene and polyamide are solids which soften and flow on heating and return to the solid state on cooling. Reinforced thermoplastics are also available.

Advantages and disadvantages of matrix materials are:

Thermosets

FOR
- Cold cure resins simplify processing
- Low-pressure moulding means cheaper tooling
- Contact moulding suitable for large mouldings and low-volume production
- Good temperature and fire resistance

AGAINST
- User must control chemical reactions and cure process
- Liquid resins have limited shelf life
- Health hazards from resin handling
- Recycling not easy
- Resins can be brittle, giving composites with low toughness

Thermoplastics

FOR
- Can be processed quickly by hot pressing or injection moulding
- Minimal knowledge of resin chemistry needed
- Available as solid pellets which are safe to handle with a long shelf life
- Ductility gives tougher composite materials
- Waste can be recycled
- Good environmental resistance

AGAINST
- High-temperature, high-pressure moulding requires expensive tooling and sensitive controls
- Expensive tooling is only cost-effective for high-volume production
- Resins soften and may burn at high temperature
- Temperature and chemical resistance varies widely

Thermosets

These resins are liquids which, when suitably initiated and catalysed, set and cure to form a hard solid. A typical formulation is given in *Table 5.6*.

Table **5.6**

Typical formulation of a thermosetting resin in parts by weight

component	proportion
polyester resin	100
accelerator	0.1 – 1.0
catalyst	1 – 5
filler	10 – 50

Note: accelerator to be premixed with resin before adding catalyst; heat will also accelerate curing.

The various stages of cure (Table 5.7) will depend upon the quantity and type of accelerator and catalyst used. These can be varied to suit the process conditions (Section 6.1) and limit the exothermic temperature rise on curing (Figure 7.22).

Table **5.7**

Curing stages of a thermosetting resin

curing time	characteristic
pot life	catalysed resin starts to thicken
gel time	moulding starts to harden
cure time	component can be demoulded
mature time	component is fully stable and hard
post-cure time	component is cured at elevated temperature

The maturing time may take hours, days or even months as the reaction proceeds towards completion. This can be monitored in the initial stages by measuring the temperature of the moulding, and subsequently by Barcol hardness (Section 7.11). Maturing can be speeded up by post-curing at elevated temperatures. This is also desirable and may be essential to obtain the optimum properties of the resin. The major types of thermosetting resins are listed in Table 5.8 and a brief description is given in Box 5.6.

Table **5.8**

Major types of thermosetting resins

type	favourable characteristics
polyester	easy to handle and cure, very versatile
epoxy	lower shrinkage than polyester good mechanical properties tough good electrical insulating properties
phenolic	stable to thermal oxidation high-temperature capability good fire retardancy
vinyl ester	excellent chemical resistance good mechanical properties
polyurethane	large family of polymers with very varied properties high reactivity integral skin with foam core possible

Box **5.6**

Types of thermosetting resins

Polyesters are unsaturated resins which are prepared by reacting an unsaturated dibasic acid with a glycol and dissolving the mixture in a reactive solvent such as styrene. On polymerizing, the styrene crosslinks, and the molecular chains form a solid copolymer of styrene and polyester. Several different acids (orthophthalic, isophthalic, terephthalic, maleic, fumaric) and glycols (ethylene glycol, neopentyl glycol, diethlyene glycol, polypropylene glycol, bisphenol A) may be used to produce resins with particular characteristics. For example, isophthalic acid provides heat resistance and bisphenol A provides chemical resistance.

A wide range of resin systems are now commercially available and resins may be obtained with superior resistance to heat, water, chemicals or flame spread, and with differing cure and viscosity characteristics. These resins form the bulk of the reinforced plastics market.

Epoxies consist typically of a linear polycondensate of epichlorohydrin and bisphenol A which is crosslinked to a solid by reaction with a hardener. These hardeners may act as catalysts or become directly involved in crosslink formation during the reaction.

Curing usually requires elevated temperatures, but curing at room temperature is also possible. Amines and acid anhydride hardeners are widely used. Depending on the resin and hardener system, the quantity of hardener required can vary from 1% to over 50% of the cured resin. Manufacturers recommend the quantity of hardener for their resin systems, and accurate measurement is essential if resins are to attain their optimum mechanical properties.

In common with polyester resins, a wide range of additives (flexibilizers and fillers) are available for particular applications. Resins may be formulated for a variety of fabrication processes, from low-viscosity liquids for filament winding to solid forms for use in prepreg sheet. Epoxies, with glass, aramid or carbon fibre reinforcement, are extensively used for advanced structural composites and currently take about two thirds of the aerospace composite market.

Vinyl esters are unsaturated esters of epoxy resins, mainly derived from reacting an epoxy with acrylic or methacrylic acid. As such they have many of the properties of epoxies with the processability of polyesters.

Phenolics are produced from a condensation of phenols and aldehydes such as formaldehyde. Use of different phenols and aldehydes, with either acid or base catalysts, leads to a range of phenolic resins with different characteristics.

Resins are available in liquid form for use with reinforcement and as prepreg sheet. Phenolics are less stable than epoxies and polyesters during storage and require more careful control of processing conditions.

Phenolics have inherently good fire-resisting properties because they char on application of heat and emit little smoke and very few toxic fumes on burning (*Figure 4.4*). Unlike other resins, they cannot be self-coloured and so a suitable coating has to be applied after fabrication. Phenolic resins are increasingly being specified for components for ▶

mass transportation, such as aircraft interior panels, rail carriage interiors and subway station fittings. Their use in these markets, where fire resistance is critical, is likely to grow considerably.

Polyimides have a niche market for high temperature (200°C plus), high-strength applications within the aerospace industry.

Polyurethanes are polymers based on the reaction product of an organic isocyanate with compounds containing a hydroxyl group. They can be used not only as engineering plastics (thermosetting or thermoplastic), but also as coatings, adhesives, sizings, sealants and as foams of varied density.

Thermoplastics

These resins are essentially composed of high molecular weight linear molecules. Unlike thermosetting resins they do not crosslink to form a polymer network on heating, but remain chemically unchanged. They melt and flow when heat and pressure are applied and re-solidify on cooling.

Fibres are generally premixed with thermoplastic resins, two main forms are currently in use:

- short fibres (usually glass or carbon) for injection moulding of small complex shaped components *(Section 6.9)*.
- continuous fibres in unidirectional tows, woven fabrics or as random mat in sheet form for hot press moulding of larger components *(Sections 5.4 and 6.3)*

For general engineering applications, polypropylene (PP) or polyamide (PA) resins are generally used, while for advanced applications, high-performance resins such as polyether-etherketone (PEEK) and polyether sulphone (PES), are now available for use with carbon and aramid fibres *(Box 5.7)* and other resins are coming onto the market.

Box 5.7

Types of thermoplastic resins

Thermoplastics used with fibre reinforcement can be divided into two main types, generally referred to as engineering plastics and advanced plastics. The principal types are:

Engineering plastics

Polypropylene (PP) is a semi-crystalline polymer belonging to the polyolefin group. It is available either as a homopolymer or as a copolymer with ethylene (5 – 15%). It is cheap and widely used.

Polyamides (PA) are a family of highly crystalline polymers commonly known as nylon. The main types are PA 6.6 and PA 6, where the numbers relate to the number of carbon atoms in the amide groups from which they are made.

Advanced plastics

Polyether sulphone (PES) is a tough amorphous resin suitable for continuous service temperatures of 180°C. It must be moulded above 300°C.

Polyether ketone (PEK) and **polyether ether ketone** (PEEK) are tough, crystalline polymers which resist environmental stress cracking. Processing is more difficult than with PP, PA and PES, and careful control of cooling rates is needed to produce the required crystallinity.

Polyphenylene sulphide (PPS) is highly crystalline, with good environmental stress cracking resistance. When ignited it generates small amounts of smoke of low toxicity. A recent variant of PPS with similar mechanical properties is polyaryl sulphone (PAS), which is available in both semi-crystalline and amorphous forms.

Modified polyimides such as polyamide-imide (PAI) and polyether-imide (PEI) have even higher temperature capability.

Some properties and characteristics of the main thermoplastic resins, described in *Box 5.7*, are listed in *Table 5.9*. Suppliers' literature or computer programs can be used to select the type and grade of resin and reinforcement.

Table **5.9**

Major types of engineering thermoplastic resins

type	melt temperature (°C)	operating temperature (°C)	favourable characteristic
PP	240	110	resistant to water and chemicals
PA	280	140	good abrasion and chemical resistance absorbs small amounts of water
PES	300	180	good chemical resistance
PEI	400	170	good chemical resistance good fire retardancy
PAI	355	230	excellent chemical resistance
PPS		220	good fire retardancy
PEEK	390	250	good toughness, difficult to process

Properties

A range of physical properties is listed in *Table 5.10* and mechanical properties are shown in *Figure 5.7* for some typical resins. Comparable values for the fibres are given in *Table 5.2* and *Figure 5.1*, from which it can be seen how much stiffer and stronger the fibres are than the resin.

Table **5.10**

Typical resin properties

property	polyester	epoxy	phenolic	PP	PA	PEEK
density (Mg/m^3)	1.3	1.2	1.2	0.9	1.1	1.3
thermal expansion (10^{-6}/K)	100	60	80	100	100	50
thermal conductivity (W/m.K)	0.2	0.1	0.2	0.1	0.2	
in-mould shrinkage (%)	4 – 8	1 – 2	0.5 – 1	2	2	
maximum service temperature (°C)	120	150	250	110	140	250
impact strength (notched Izod)(J/m)	15	20	–	100	80	80

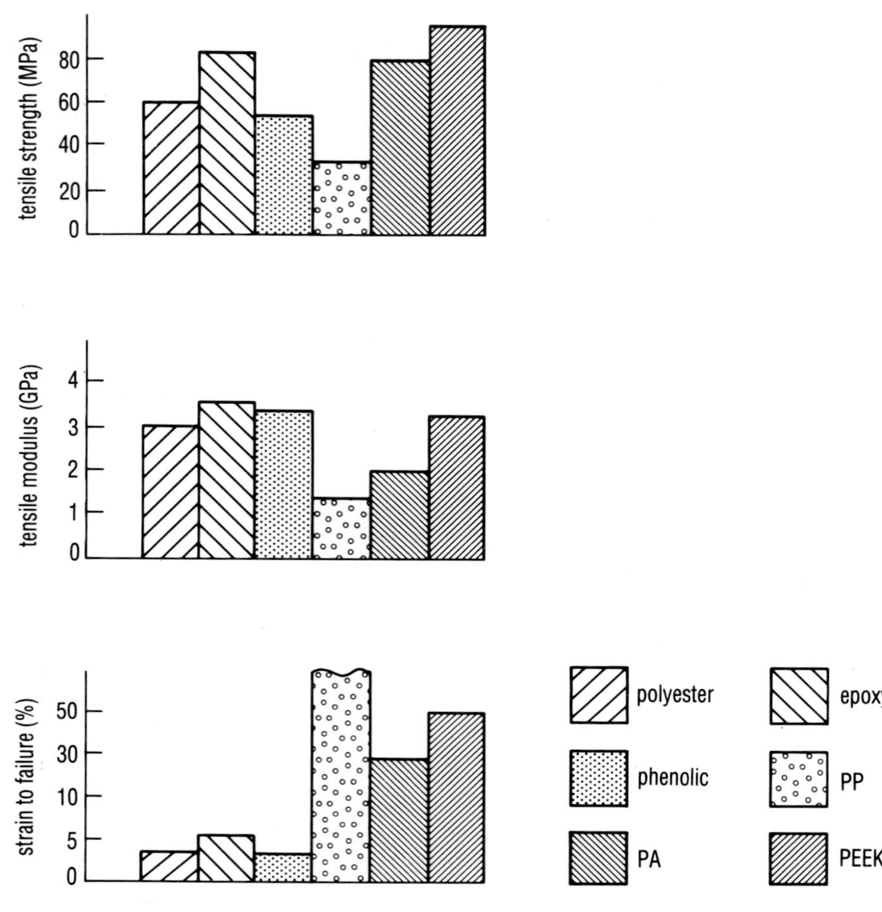

Figure 5.7

Tensile modulus, ultimate strength and strain to failure for some resins. These values can be compared with those of the fibres in Figure 5.1

5.4 **Compounds and intermediate products**

Compounds are available in various forms in which the resin, fibres and possibly fillers are premixed before moulding *(Table 5.11)*. Suppliers will indicate the process conditions needed to achieve particular properties, though this should always be checked in practice.

Moulding compounds are available with both thermosetting and thermoplastic resins. Most compounds have been developed for a particular machine-controlled fabrication method, such as hot press moulding *(Section 6.3)* or injection moulding *(Section 6.9)*. Use of these compounds permits large production runs with good quality control of both material content and component fabrication, with reduced health hazards from styrene emissions.

The composition and use of such materials are described in Box 5.8.

type	form	reinforcement length (mm)	thickness (mm)	process
IMC	compound	1 – 2	–	IM
DMC	compound	3 – 12	–	PRM, IM
SMC	sheet	20 – 50*	–	PRM
GMT	sheet	random mat	1 – 5	PRM
TSC	sheet	woven	0.1 – 0.25	PRM
prepreg	sheet, tape	continuous	0.1 – 1	PRM

Notes: *IM* *injection moulding* (Section 6.9); *IMC* *injection moulding compound*; *SMC* *sheet moulding compound*; *TSC* *thermoplastic sheet compound*; *PRM* *press moulding* (Section 6.3); *DMC* *dough moulding compound*; *GMT* *glass mat thermoplastic* (Section 5.3)

* *continuous reinforcement forms of SMC also available* (Figure 6.6)

Table **5.11**

Major types of moulding compounds

Box **5.8**

Types of moulding compounds

Injection moulding compounds

IMCs are injection moulding materials with chopped fibres added to enhance the mechanical properties of traditional injection moulding thermoplastics such as PP and PA, or thermosetting resins such as epoxies, polyesters or phenolics.

They are used for high-volume production of small, intricate mouldings, usually less than 1 kg in weight, such as gear wheels, door handles, pumps, and fans. The chopped fibres are usually incorporated into the thermoplastic at about 10 – 25% by volume. Injection moulding breaks the fibres down, and typical fibre lengths in a moulding are usually less than 0.25 mm unless special precautions are taken.

Thermosetting moulding compounds, such as chopped glass fibres in a polyurethane matrix, are also available and because of their low viscosities are suitable for the injection of panel components. With such a material the composite is formed by chemical reaction in the mould by combining two liquid intermediates, hence the term reinforced reaction injection moulding (RRIM) *(Box 6.9)*.

Fibre orientation distribution in the moulding is influenced by the flow pattern in the mould cavity during processing. Fibres tend to align in the flow direction in converging and shear flows, while in diverging flow they are aligned normal to the flow.

It follows that a moulding may have regions with quite different fibre orientation. The mechanical properties of such mouldings can be considerably worse than those of long-fibre composites, which makes data for design purposes unreliable.

Dough moulding compounds

DMCs, sometimes called bulk moulding compounds or BMCs, contain catalysed polyester resin, up to 50% mineral filler and 15 – 25% short glass fibres by volume. The usual filler is chalk, but some resins are chemically thickened with alkali metal oxides.

Having shorter fibres and less glass than SMC enables DMC to flow more easily, so it is used to make small, intricate articles by hot press or injection moulding. Strength is lower than for SMC. ▶

DMC and SMC materials were developed with glass fibre reinforcement for high-volume applications such as building panels, vehicle body panels and trim, and engine components.

Sheet moulding compounds

SMCs usually consist of glass fibres reinforcing a mixture of catalysed polyester resin and a mineral filler, such as particulate chalk, limestone or clay. The filler thickens the resin sufficiently for uncured sheets of SMC to be stored, handled and cut to the correct size for moulding.

SMC usually contains chopped glass strands randomly orientated in the plane of the sheet with up to 35% of fibre and 30 – 50% of filler by volume. High-performance SMC materials may also contain aligned chopped glass fibres or continuous fibres at up to 70% by weight, e.g. XMC *(Section 6.3)*.

Glass mat thermoplastics

GMTs comprise a random glass mat embedded in a thermoplastic matrix. Typical resins are engineering plastics such as polypropylene or nylon, with 30% of fibre by volume. They are suitable for high-volume production of panel-type components such as hot pressed or stamped engine valve covers or car body components.

Thermoplastic sheet compounds

TSCs consist of continuous fibre sheets or tapes impregnated with thermoplastic resin. They have much better and more reliable mechanical properties than GMT. Carbon and aramid as well as glass can be used as reinforcement.

For high-performance applications, unidirectional tows or a finely woven fabric are used as the reinforcement, typically with nylon or PEEK, with 40 – 60% fibre by volume.

These materials can be regarded as thermoplastic prepregs without the cold storage requirements and with long shelf lives.

Prepregs

Prepregs consist of sheets or tapes of fibre reinforcement preimpregnated with catalysed resin and ready for autoclave moulding or vacuum bagging *(Section 6.8)*. They are formed using all the major fibres for high-performance applications such as racing car body shells, aircraft wing panels and helicopter rotor blades.

They require cold storage in a refrigerator and even then have a shelf life measured in weeks. Once exposed to ambient temperatures, shelf life is likely to be only a few hours.

The reinforcement is usually unidirectional tows or fabric with epoxide rather than polyester resin, and component thickness is achieved by laminating sheets together in the mould. Up to 100 layers may be used in an aircraft wing. Fibre content is high, typically 50 – 80% by weight. Accurate alignment of fibres is possible with prepregs because there is minimal flow during moulding.

5.5 Composites

The properties of a fibre reinforced composite are a mixture of the properties of its constituents. Some properties depend upon the fibre and some upon the resin *(Figures 2.8 and 2.9)*. This chapter considers one property – the response to a tensile load – in some detail as this mechanical property tends to be most important for the designer.

Loading in tension

On loading a FRP material in a short-term tensile test, the material responds in an approximately linear elastic fashion to failure, which occurs by a brittle fracture at a strain of 1.5 – 2.5%. This is illustrated for a chopped strand mat/polyester composite in *Figure 5.8*.

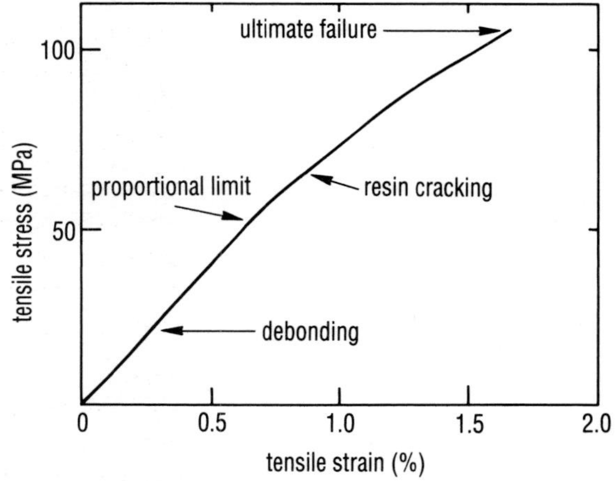

Figure **5.8**

Typical stress- strain curve for a chopped strand mat glass laminate in polyester resin (Johnson)

The short-term mechanical properties can therefore be characterized by an elastic modulus and a failure stress. A more detailed study of failure shows that there is progressive damage to the material at stresses below the ultimate.

The first sign of damage is transverse fibre debonding through separation between the resin and fibres perpendicular to the load direction. This occurs typically at 0.3% strain and a stress level of about 30% of the tensile strength.

Under increased loading, the debonds initiate resin cracks and, as these cracks spread, more of the load is transferred to the fibres which eventually fracture or pull out at ultimate failure. Resin cracking depends on the toughness of the resin, but typically occurs at approximately 50 – 70% of the tensile strength and is observed as crazing and by a change of slope in the stress/strain curve.

Design implications

Ultimate failure stress data are usually given in tables such as *Table 5.12*, although for many applications failure data based on alternative criteria such as resin cracking or fibre debonding may be necessary. Some data of this type are available, or alternatively a suitable design factor may be applied to the ultimate strength data. With a knowledge of the appropriate moduli and strengths, elastic design formulae may be used for stress and deformation analysis of fibre reinforced plastics under short-term loads *(Section 3.3)*.

For random mat and SMC reinforcement, conventional isotropic design formulae may be used, while for woven and knitted fabrics or yarn emplacement, account must be taken of the anisotropy in stiffness and strength properties arising from fibre orientation *(Figure 3.8)*.

Design data

It is not usually possible to quote a single value of modulus and strength because of the heterogeneous structure of the material. Thus, in addition to the effects of anisotropy, mechanical properties also depend on the quantity and type of fibre reinforcement and on the fabrication method used. Bounds for properties which allow for the influence of these various parameters are discussed elsewhere *(Section 4.7)*. Here, property data is presented as typical single point values as an aid to the designer selecting materials.

It should be remembered that data quoted in this guide is representative of a spectrum of values for a material that will vary with details of fibre proportion and alignment, resin properties and processing method.

Table **5.12**

property	DMC	SMC	CSM	WR	UD
glass content (% by weight)	20	30	30	50	70
modulus (GPa)					
tensile	9	13	7.7	16	42 [12]
flexural	8	11	6.3	13	
shear	3	3	3	4	5
strength (MPa)					
tensile	45	85	95	250	750 [50]
flexural	100	180	170	290	1200
in-plane shear			80	95	65
interlaminar shear	15	25	20	40	
impact strength (unnotched Izod)(kJ/m2)	20 – 40	50 – 75	40 – 80	100 – 200	

Notes:
CSM chopped strand mat
WR woven roving (fabric)
UD unidirectional fabric
[ ] properties transverse to axis of main reinforcement
resins are all polyester except UD reinforcement/epoxy

Data may be given in the form either of fibre volume fraction or fibre weight fraction. The conversion formula is

$$V_f = W_f \rho_m / [W_f \rho_m + W_m \rho_f]$$

Where V is the volume fraction, W the weight fraction, ρ the density and subscripts $_f$ and $_m$ denote fibre and matrix.

For example, the WR composite in *Table 5.12* has a fibre weight fraction of 0.50. So its fibre volume fraction would be:

$$V_f = 0.5 \times 1.2 / [0.5 \times 1.2 + 0.50 \times 2.5]$$
$$= 0.32$$

Mechanical properties for the main glass fibre reinforced thermosets are given in *Table 5.12* and physical properties in *Table 5.13*. These contain data on DMC and SMC moulding compounds, together with comparable properties for the three main types of fibre reinforcement. Corresponding mechanical properties for high-performance thermosetting composites with aramid and carbon fibre reinforcement are given in *Table 5.14*.

Table **5.13**

Typical physical properties of GRP materials

property	CSM, SMC polyester	WR fabric polyester	UD epoxy
density (Mg/m^3)	1.3 – 1.6 (CSM) 1.6 – 1.9 (SMC)	1.5 – 1.9	1.8 – 2.0
thermal expansion coefficient (10^{-6}/K)	18 – 35	10 – 16	5 – 15
thermal conductivity (W/m.K)	0.16 – 0.26	0.2	0.3
specific heat (J/kg.K)	1200 – 1400		950
maximum heat distortion temperature (°C)	175	250	300
dielectric strength (kV/mm)	9 – 12	13 – 16	
permittivity at 1MHz	4.3 – 4.7	4.1 – 5.2	
dry insulation resistance (M ohms)	10^6	10^6	
power factor at 1MHz	0.015	0.016	

Note: WR fabric is balanced, warp and weft.

Table **5.14**

Typical mechanical properties of aramid and carbon fibre reinforced epoxy resins

property	UD aramid epoxy	woven aramid epoxy	HM-UD carbon epoxy	HS-UD carbon epoxy	woven carbon epoxy
fibre content (% by volume)	60	60	60	60	65
density (Mg/m^3)	1.4	1.4	1.6	1.5	1.6
modulus (GPa)					
tensile	90	37	200 [7]	137 [10]	70
flexural	80	30	190	130	67
strength (MPa)					
tensile	1400	550	1150 [35]	1900 [50]	510
flexural	670	450	1100	1800	810
interlaminar shear	69	35	70	90	62

Notes:
- *HM* — *high-modulus carbon fibre*
- *HS* — *high-strength carbon fibre*
- *UD* — *unidirectional tows*
- *woven* — *balanced fabric (warp/weft)*
- *[ ]* — *properties transverse to axis of main reinforcement*

Some limited data on thermoplastic composites are given in *Table 5.15*. The data for the glass filled injection moulding materials are typical of injection moulded test bars and are probably higher than the properties attained in other mouldings with more complex flow patterns.

Table 5.15

Typical mechanical properties of fibre reinforced thermoplastic composites

	glass mat PP	glass PP	glass PA	glass PPS	carbon PEEK
fibre content (% by volume)	30	20	20	70	60
density (Mg/m^3)	1.2	1.1	1.5	2.0	1.7
modulus (GPa)					
tensile	5.5	5.7	10	50	134 [9]
flexural	4.5			44	121 [9]
strength (MPa)					
tensile	85	103	210	910	2130 [80]
flexural	140			1160	1880 [237]
interlaminar shear					100

Notes:
PP and PA — short fibres
PPS and PEEK — aligned long fibres
glass mat/PP — a glass mat thermoplastic
[ ] — properties transverse to axis of main reinforcement

Typical mechanical properties for pultruded sections are given in Table 5.16 and further information can be obtained from manufacturers of finished products of this type.

Table 5.16

Typical mechanical properties for pultruded sections with glass fibre reinforcement (fibre force)

property	random mat/rovings polyester	rovings polyester
flexural strength (MPa)	200 [70]	700
flexural modulus (GPa)	9 [4]	40
compressive strength (MPa)	140 [70]	400
density (Mg/m3)	1.6	1.9

Notes:
[ ] = transverse measurements
test methods: ASTM
rovings fully aligned

5.6 **Reference information**

General reference

Concise Encyclopedia of Composite Materials, A Kelly (ed), Pergamon, Oxford, 1989 – comprehensive set of short articles covering all aspects of materials.

A Guide to High Performance Plastic Composites, BPF, London, 1980.

Engineering Design Properties of GRP, A F Johnson, BPF, London 1984 – data collated and analysed from a wide variety of sources – very useful guide.

Engineering Design in Plastics – data and applications guide, G H West, PRI, London, 1986 – contains a list of short-term data for design as well as simple design formulae.

Engineering Properties of Thermoplastics, R M Ogorkiewicz (ed), Wiley, New York, 1970 – unfilled resins only, a comprehensive description of thermoplastic properties from one main material supplier.

Design Manual of Engineered Composite Profiles, J A Quinn, Fibreforce, Runcorn, 1988 – useful tabulation of properties of pultruded sections.

Engineering Plastics, ASM International, Metals Park, 1987 – handbook on engineered materials, standard reference work.

FRP Technology, R G Weatherhead, Applied Science Publishers, London, 1980 – standard work, very comprehensive on resins.

Elsevier Materials Selector, N A Waterman and M F Ashby (ed), Elsevier, Barking, 1991 – vol 3 composites – comprehensive set of information.

Specific references

5.2 **Fabrics**

Knitted Fabrics – guide to technical textiles, S Raz, Karl Mayer, Obertshausen, 1988

Polyester Handbook, Scott Bader, Woolaston, 1986.

'Bring in the reinforcements', E Taylor in *Advanced Composites Engineering,* January 1990 – description of most recent developments in weaving.

Data sources of tables

Tabular data has been obtained from a number of sources, primarily those listed in the general references above, with some additional data and cross-checks from suppliers' literature.

Materials properties on data bases

An increasing amount of information is stored in computer databases which may be consulted either by linking up with a mainframe computer or by using a disk with a personal computer. Almost all are completely searchable in the sense that one may begin with a material, a property level, or even a cost. Some databases contain more information on properties than others.

Some suppliers, notably the manufacturers of thermoplastic resins such as ICI and Bayer, will supply a disk free of charge, but these are usually restricted to the suppliers' own materials. Initially, designers should consult a reference library, such as the Institution of Mechanical Engineers or an advice point such as the Materials Information Service of the Design Council, which will be able to provide guidance as to which database(s) to consult and may be able to undertake the search as well:

Institution of Mechanical Engineers (loc cit) +44 071-222 7899
Design Council, 28 Haymarket, London SW1Y 4SU +44 071-839 8000

The databases developed by RAPRA Technology (loc cit), such as PLASCAMS and CHEMRES, are a good starting point as these are available with a suppliers list for each of the main European countries. In addition, they contain an assessment (value-judgement rating) of all the listed materials against 67 searchable quantities.

For a database that compares materials other than plastics, a suitable starting point might be either MATUS or PERITUS; further details of these systems can be obtained from the Materials Information Service of the Design Council.

ASM database sets include information on nine reinforcement materials and 17 resin matrix materials (1992), as part of their composite data sets. European agents are Comline Ltd, Blakes House, 98 Ickleford Road, Hitchin, Hertfordshire SGS 1TL, UK +44 0462-453 211.

M/Vision from PDA Engineering 2975 Redhill Avenue, Costa Mesa, California, is a workstation-based materials databank system containing both graphical and tabulated data on a range of materials. For composites it contains validated data from the US military handbook for composites MIL-HDBK-17A, together with the databank PMC-90, which was the result of a characterisation project for polymer composites carried out at the US Air Force Materials Laboratory.

Design Data for Reinforced Plastics

This book, published by Chapman & Hall as a companion to this text, provides a design data set covering the common reinforcement and matrix combinations. Information has been gleaned either from material suppliers or from scrutiny of bibliographic base compiled by RAPRA Technology (loc cit).

Footnote

Chapter 5 was written by Alastair Johnson of the German Aerospace Establishment (DLR), Institute for Structures and Design, Stuttgart, Germany.

Manufacturing considerations

chapter 6

summary

When using reinforced plastics, the material is manufactured at the same time as the component, so selecting a suitable manufacturing process is crucial for the success of the design. In this chapter, a table is used to select the appropriate production process. Each process is then described in detail so that the initial choice can be checked against a summary table of advantages and disadvantages.

Introduction

The use and acceptance of fibre reinforced plastic materials for engineering components and load-bearing structures has grown at a steady rate over the past two to three decades. Initial fears and resistances have been overcome with each successful demonstration that these materials could fully justify the claims made for them, and that their attractive properties could be realized.

Most, if not all, of the early hi-tech applications were in the aerospace industries. Manufacturing methods, design tolerances and levels of quality control reflected the particular requirements and performance specifications of that sector with regard to both time and cost.

More recently, growing awareness of and confidence in this class of materials has led to their appearance in less exotic components and structures. Their increasing use for mass production rather than in one-off or short-run applications has resulted in demands for reduced production times, improved quality control and greater repeatability.

This has led in turn to the development and improvement of existing processing methods as well as the introduction of new ones. It is therefore important for the designer to be fully aware of the interaction and inter-relationship between the various composite materials, their properties, and the processing routes used in making them.

It cannot be stressed strongly enough that the designer in composites has to consider not only the choice of a suitable material but also the manufacturing method. Selecting a set of desirable material properties from a table in a supplier's handbook or publicity sheet is no guarantee that these properties can be attained in a given application. Certain property levels, component geometries and features are more appropriate for some processing methods than others, and some features are inappropriate with some methods.

There is no rigid or definitive set of rules for the manufacture of components in fibre reinforced plastics. Initial restrictions to the use of a particular method may be avoided by:

- changes to the original material specification.
- modifications to the shape of the component *(Section 3.1)*.
- incorporating adjacent parts or functions *(Section 1.2)*.

However, a set of guidelines can be offered that allows an initial assessment and an indication of the potential or feasibility of the various processing routes *(Table 6.3)*. Each designer will develop this table with experience, introducing a finer ranking system, better tuned to the demands of particular jobs or industries.

6.1 **Process selection**

Production processes, as well as the factors involved in selecting a particular process, have already been briefly described *(Section 2.6)*, and these are listed in *Table 6.1*. Factors that will affect and should influence the final process selection include the component's size and whether it is cosmetic or structural, one-off or mass produced, made in-house or subcontracted.

A guide to process selection is set out in *Tables 6.1*, *6.2* and *6.3*. Decisions should be confirmed subsequently by consulting the appropriate section in this chapter.

6.1 **Process selection**

process	abbreviation
contact moulding	COM
press moulding	PRM
resin transfer moulding	RTM
filament winding	FIL
centrifugal moulding	CEM
pultrusion	PUL
prepreg moulding	PPM
injection moulding	IM

Table **6.1**

Types of manufacturing processes

Notes:
- *Reaction injection moulding involving the use of reinforcement (RRIM) is similar to injection moulding (IM)*
- *Prepreg moulding covers use of both vacuum (vacuum bag) and pressure (autoclave)*

Table **6.2**

Typical process parameters

process parameter	unit	designation of descriptors and values		
size	metres	small (<1)	medium (1 – 3)	large (>3)
shape	–	constant	cylindrical	any
numbers	per day	small (<3)	medium (3 – 50)	large (>50)
labour intensity	minutes per item	low (<1)	medium (1 – 20)	high (>20)
capital intensity	£,000 per item	low (<3)	medium (3 – 30)	high (>30)
reinforcement length	mm	chop (<2) continuous	short (2 – 25)	long (25 – 150)
possible use of compounds	–			
number of good surfaces	–	discrete number, dependent upon mould geometry and process		

Note:
Compounds, refer to Section 5.4

Table **6.3**

Process selection

process	COM	PRM	RTM	FIL	CEM	PUL	PPM	IM
size	large	med	med	large	large	med	med	small
shape	any	any	any	cyl	cyl	const	any	any
number	small	med	med	med	med	large	small	large
labour	high	med	med	med	med	low	high	low
capital	low	med	med	med	high	med	med	high
reinforcement	any	short	any	cont	long	cont	cont	chop
compounds	no	yes	no	no	no	no	yes	yes
surfaces	one	all	all	three	two	all	all	all

Notes:
cyl – cylindrical section
cont – continuous reinforcement
const – constant section
med – medium

6.2 **Contact moulding (COM)**

Contact moulding is a process in which reinforcement and resin are sequentially emplaced into an open mould, and it is probably the best known and most widely used method for producing fibre-reinforced plastic parts. It is commonly used for short runs or one-offs, and because it is very labour intensive and time-consuming, it results in long production times. More recently, the basic process has been adapted for longer production runs on simple parts using automatic or semi-automatic equipment.

There is no limit to the size or complexity of structures that can be produced by this method. A wide range and variety of parts have been made, from small, non-structural, aesthetic articles up to large structures such as the shells of wing blades *(Box 1.3)* and complete hulls for the Royal Navy 'Wilton' Class mine-sweepers.

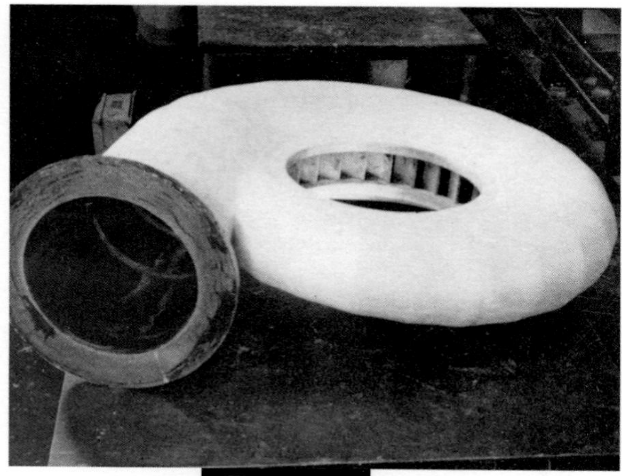

Figure **6.1**

Spiral pump (NEL)

Mouldings made by this process are generally self-coloured and possess only one smooth finished surface. Whether this is on the top or bottom, inner or outer surface will depend on whether a male or a female moulding tool is used. Textured finishes are also possible on this primary surface of the moulding.

Ideally, the part or structure being made should have a draft angle to facilitate removal from the mould, but re-entrant shapes and forms can be achieved by using a cored or collapsible mould, as in the prototype pump casing *(Figure 6.1)*.

Moulds

Only one mould (or moulding tool) is needed to produce the moulding and it can be made from a variety of materials including wood, plaster, GRP or metal. These materials, separately or in combination, are used to provide a strong, rigid tool with a smooth skin stiffened by ribs or backfilling.

These working moulds are normally made from a master pattern or moulding *(Figure 6.2)*, which allows for good repeatability over a production run.

Alternatively, thermosetting resin systems have been developed that allow the mould to be cast directly from a master pattern. These commercially available compounds are ideal for longer runs, whereas wood or plaster suit the shorter runs or one-off mouldings.

Materials

All types of reinforcement can be used, from common fibres such as glass and jute through to more exotic materials like carbon. They can be in many different forms including mats, fabrics and continuous rovings *(Section 5.2)*.

Most fabricators use unsaturated polyester resin, which enables the component to be cured at room temperature *(Section* 5.3). Other resin systems can also be used, the most common being vinyl ester and epoxy.

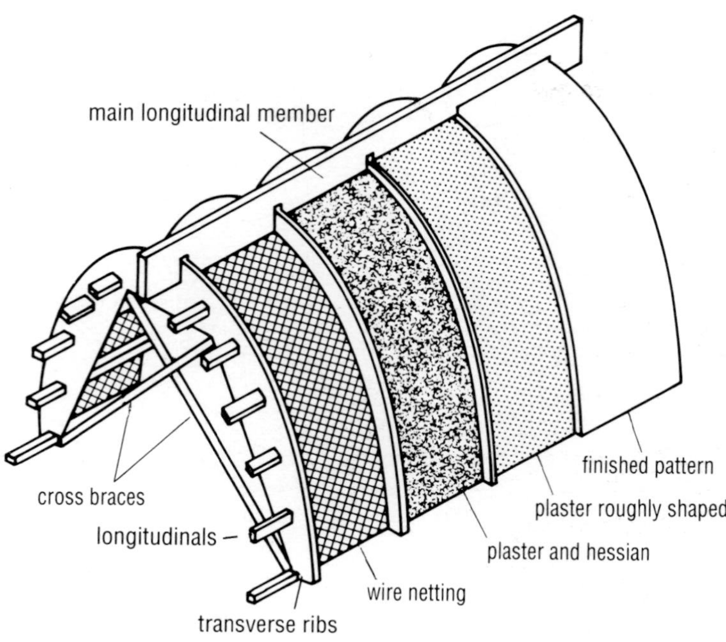

Figure **6.2**

Master pattern make-up (Scott Bader)

A variety of **additives** can be mixed in, ranging from plasticizers and flame retardants to inert fillers which can reduce weight and improve the mechanical and physical properties of the final component. Some of these are listed in *Table 6.4*.

Table **6.4**

Some additives and their uses

type	use	typical materials
colourants	colouring	pigments and dyes
flame retardants	reduce smoke, inhibit ignition	halogen or water releasing agent
plasticizers	increase flexibility of resin	polymer of lower weight
inert fillers	reduce cost, modify properties	chalk, calcium carbonate, glass spheres
foaming agents	form cellular foams	chemicals that generate inert gas

Core materials can also be used to increase the bulk and form a geometric shape, thereby adding stiffness and reducing the mass of components *(Section 3.1)*. These can be made of expanded foam, honeycomb core, paper, cardboard, wood or other similar light material.

Typical properties of composites manufactured by this process are given in *Table 5.12*.

Lay-up process

Release agents are used to prevent the part sticking to the mould. These come in various forms such as waxes, polyvinyl alcohol and specially formulated systems usually supplied in aerosol cans.

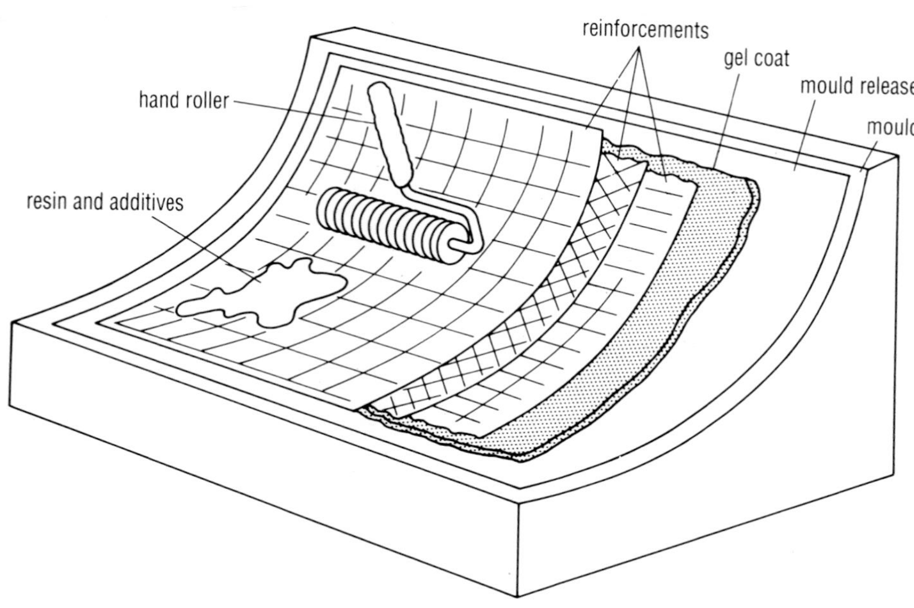

Figure **6.3**

Schematic of contact moulding (Richardson)

The process of laying up a component begins with the application of the **gel coat** to the surface of the mould *(Figure 6.3)*. This resin-rich layer (0.4 – 0.5 mm thick) provides a smooth, hard surface, prevents the fibre reinforcement appearing on the surface, and can be pigmented to colour the part. The gel coat may be applied either by a spray-gun or by hand using a paint brush. This coat may also in some cases be reinforced with a very fine glass tissue or veil.

The **fibre** reinforcements are then laid on over the gel coat surface by hand and the resin is applied to the reinforcement by brush or roller, layer by layer, building up the moulding to the correct thickness. During this stage the material is continually worked and rolled to ensure proper wet-out of the fibres, expulsion of air and consolidation of the material.

Typical glass to resin ratios of 15 – 20% by volume can be achieved using random mats and up to 45% using fine fabrics. The moulding is then left until the resin has fully cured. The time taken for curing depends on the resin, catalyst, temperature and thickness of the moulding.

Spray moulding

For larger components (say a few square metres), and where the reinforcement fraction need not be high, the process can be semi-automated by using a special type of spray gun. The reinforcement is chopped and an air stream is used to feed the mixture of chopped fibres and catalysed resin onto a mould. As with hand lay-up, consolidation must be done with a roller.

Summary: contact moulding

advantages	disadvantages
all shapes and sizes	one moulded surface only
accepts most reinforcements	hand compaction only
low tooling cost	labour intensive
inserts possible	high volatile emission
ideal for prototypes	quality control difficult
	slow production rate

Table **6.5**

Contact moulding: advantages and disadvantages

6.3 **Press moulding (PRM)**

Press moulding involves the use of a press and a moulding compound or intermediate material in which the resin, reinforcement, catalyst (if any) and additives are already premixed in the optimum proportions.

Hot compression moulding is the most common form of moulding process involving a press, heated platens and matched tooling. It uses a thermoset moulding compound like SMC or BMC *(Box 5.8)*. Cold press moulding uses pressure alone to impregnate reinforcement laid up in a mould with catalysed resin, which subsequently cures at room temperature.

A similar hot press moulding process can be used with reinforced thermoplastic sheet compounds such as GMT and TSC, in which case the process tends to be called thermoforming or stamping.

A wide range of parts can be produced by these processes, from electrical housings, car body panels, gas and electricity meter boxes to car bumpers. There is now a move towards using hot press moulding to produce fully structural load-bearing components such as car road wheels *(Figure 6.4, Section 8.2)*.

Figure **6.4**

Plastic wheel fitted to a standard car (moulded at NEL)

Processing

Hot press moulding involves squeezing and flowing a charge or preform of moulding compound within a mould cavity which forms the finished shape of the part being made *(Figure 6.5)*.

The mould tool is a matched metal assembly made from aluminium, cast iron or steel which is heated by steam, electricity or hot oil to temperatures of between 140 – 160°C. The heat transferred from the mould surface to the mould material initiates the curing system. Curing times vary according to the resin system used and with component thickness. A typical value of 1 minute for each 2 mm section thickness based upon the thickest section may be used. Typical cycle times are from 2 – 4 minutes. Processing equipment is normally hydraulic, with controls to govern and monitor the moulding cycle within close limits.

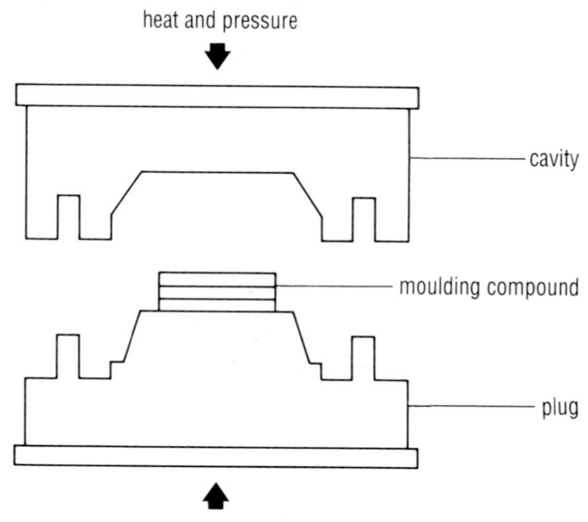

Figure **6.5**

Schematic of moulding press and matched tooling. Pressures are typically in the range of 65 –100 bar (OCF)

Tooling costs tend to be relatively high, so the process is only suitable for long production runs, although this depends on the complexity of the tool and whether it has removable cores. Multi-cavity tools can be used for simpler mouldings, increasing production speeds and reducing overall component costs.

If the moulding compound is in sheet form, then the sheets are precut to the appropriate size and are generally preheated before being loaded into the mould. This can be done with both thermoplastic and thermoset moulding compounds.

Materials

The materials used for hot press moulding and thermostamping are specially prepared and formulated compounds and incorporate a mixture of resin, reinforcement and fillers *(Box 5.8).*

Several different types of resin systems can be used: among the thermosets, primarily unsaturated polyester, but more recently vinyl ester and phenolic resins; among the thermoplastics, polypropylene, nylon or, for more demanding applications, PEEK.

The reinforcement is generally glass, randomly orientated and with a fibre length up to 25 mm, but various types of directional reinforcement are also available *(Figure 6.6).*

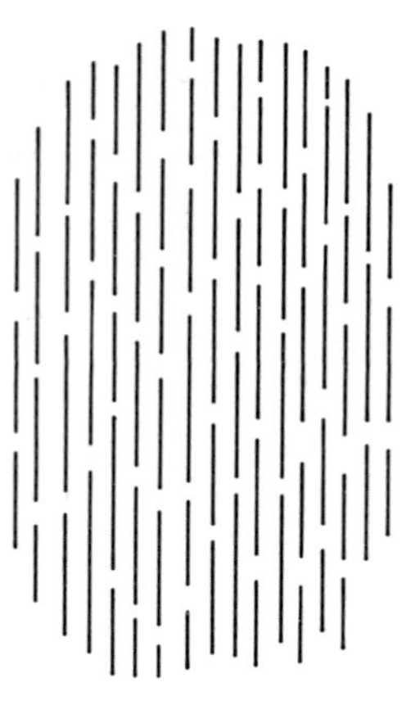

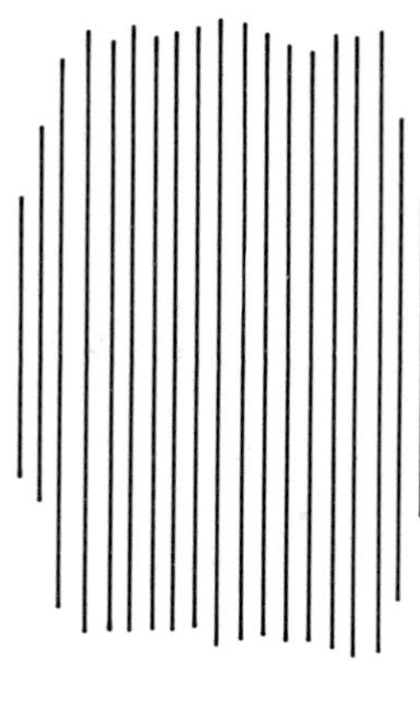

Figure **6.6**

Types of aligned reinforcement: (a) random alignment such as SMC and GMT; (b) aligned such as XMC or TSC (c) continuous aligned rovings such as prepreg (Table 5.11) *(OCF)*

Fibre loadings of up to 60% by volume can be achieved, but high fibre fractions require a high degree of fibre alignment to aid packing of the reinforcing fibres. Typical properties are given in *Tables 5.12* and *5.15.*

Summary: press moulding

advantages	disadvantages
fast cycle times	some post-machining and trimming may be necessary
near neat shape forming	difficult to obtain a first-class finish
process automation	high tooling costs
complex shapes	
parts integration	
self colouring	
structural integrity	

Table 6.6

Press moulding: advantages and disadvantages

6.4 Resin transfer moulding (RTM)

This technique is an established process, which has become more popular in recent years as a result of interest by the automotive and aerospace industries.

It is inherently very simple, involving the transfer of resin into a closed mould into which the reinforcement has already been placed. The use of a closed mould enables good tolerances to be achieved (within 1 part in a thousand) without the need for machining and enables emissions of volatiles to be controlled.

Examples of components made by this technique include manhole covers, compressor casings, car doors and side panels *(Box 1.1)*, and propeller blades (Figure 6.7).

The technique is very versatile and can cover a wide range of shapes and sizes and use a wide variety of reinforcements. Inserts are also possible.

Figure 6.7

RTM moulded part (moulded at NEL)

Process

Reinforcement is cut out using a template and a knife or scissors. For large quantities a pattern cutter and a roller press should be considered. Reinforcement can then be inserted into the mould by laying up individual layers of fabric with the correct orientation.

The lay up process can be facilitated if the fabric has a small amount of thermoplastic binder on the surface which can be softened by heating. In this way the set of reinforcements can be preformed to the required shape and trimmed to size before insertion into the mould. A number of methods of preforming exist and others are being developed. The choice depends on the number of preforms required, the investment available and the performance required of the component *(Box 6.1)*.

Box **6.1**

Methods of preforming reinforcement

Preforms

Reinforcement can be preformed if a small amount of thermoplastic binder (2 – 5%) is present on the surface of a fabric. Once softened by heat, the fabric can be set to shape and allowed to cool.

Random fibre preforms

For components where shape and cost are most important, a random mat can be used. Preforms can then be produced by hand, using the open mould as the lay-up tool and a warm-air gun to soften the binder. The preform is built up to the required thickness by cutting and placing any additional reinforcing strips onto the preceding layer.

Another technique uses lightweight tooling onto which the chopped fibres are sprayed together with a suitable binder material. Heating or drying consolidates the preform which is then transferred to the mould for processing. Plenum chambers are also used for consolidating the chopped fibres if a higher production rate is required.

High performance preforms

To obtain a high volume fraction and thus improve the properties of the component, matched lightweight tooling is used to compress the reinforcement under pressure. Continuous fibres in the form of woven or non-woven fabrics on their own or in combination with random mat can be aligned to lay up reinforcement along the principal stress directions.

Binding methods.

Various methods can be used to bind the layers including:

- spraying a liquid-based thermoplastic binder onto the fabric and bringing the layers together before the solvent evaporates.
- using a double-sided adhesive glass tape which holds the layers together (Admat).
- using an open weave fibril coated with a compatible binder which, when heated, will flow onto the fabric and hold the preform together when cooled (Crenette).
- using a fabric coated with a thermoplastic powder binder already on one surface.

It is also possible to incorporate such items as foam cores, inserts, fasteners and pre-shaped stiffeners into the preform.

To achieve larger production quantities and greater reproducibility, a moulding line can be set up simply to produce the preforms.

Preforms are ideal for improving quality control on the reinforcement, as the mass, orientation and uniformity can be checked before the resin is added.

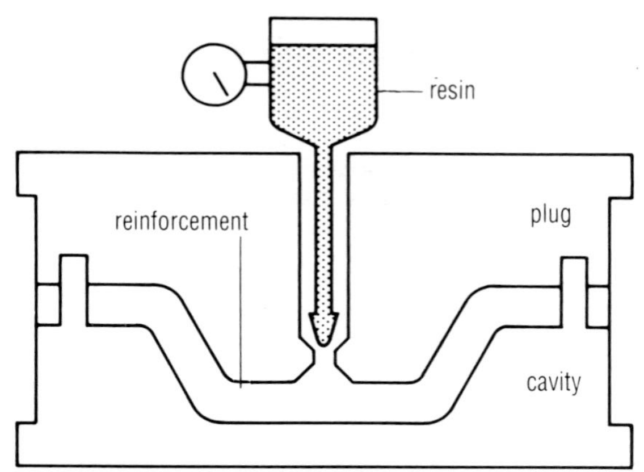

Figure **6.8**

Typical RTM tool in which reinforcement is laid into mould and resin subsequently injected (OCF)

Resin is transferred into the mould with a variety of equipment, from very simple to expensive and complex. Several variations on the same theme can be used to incorporate either pressure or vacuum *(Box 6.2)*.

Developments into the use of robotic handling, cutting and placement of preforms are under way to achieve the volumes and production speeds required by the automotive industry.

Tooling

An essential feature of this process is that the resin drives the air ahead of it and out of the mould *(Figure 6.8)*. The air provides a back pressure which ensures that the resin wets out the reinforcement as it flows past. A variation on this theme, developed by Lotus in the UK, uses a partial vacuum to draw resin into the mould. This process has been used for cars and boats. The flexibility of the process and its ability to cover a wide range of component volumes also extends to the tooling. For prototype parts and small volume runs a simple cast tool may be all that is necessary. For larger components, lightweight tooling is usually fabricated from fibre reinforced plastics.

Box 6.2

Methods of injecting resin into a mould

Low to medium volume production

The simplest method is to use air pressure and a resin gun to mix resin that has been premixed with either catalyst or accelerator and inject it into a mould.

Another machine (Hypaject), also uses a premixed resin system which is drawn into the transfer chamber by an induced vacuum (*Figure 6.9*). A simple switch allows air pressure to act on the surface of the resin which in turn forces the resin into the tool through a control diaphragm. The flow is regulated by adjusting the air pressure.

This equipment is ideal for laboratory or prototype work due to the ease with which resin systems can be changed, and recent developments to increase its capacity have allowed it to be considered for industrial production. ▶

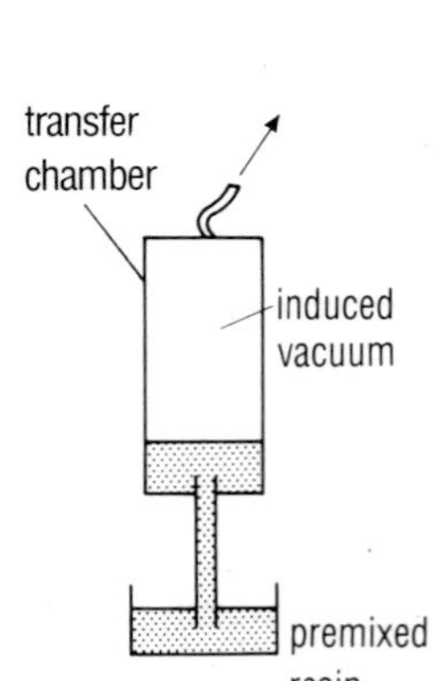

Figure 6.9

Principle of the Hypaject. Resin is drawn into a transfer chamber by a partial vacuum and then ejected into a mould using compressed air

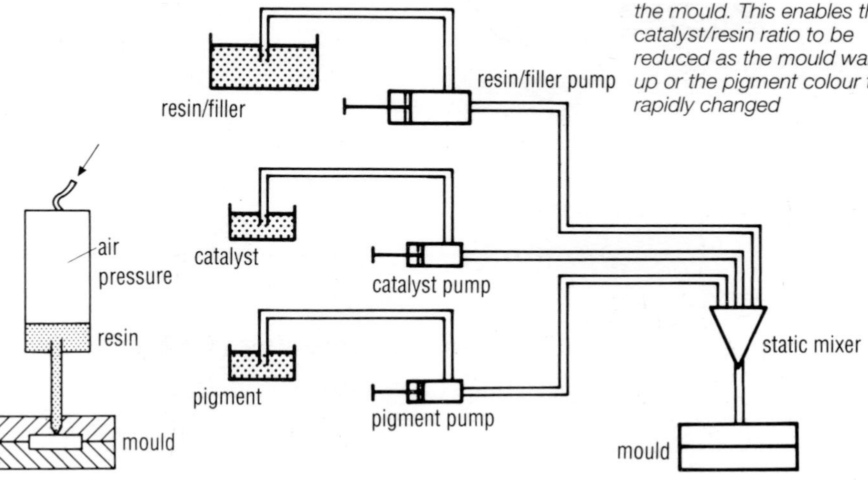

Figure 6.10

The use of separate displacement pumps enables the resin, catalyst and pigment to be mixed prior to entering the mould. This enables the catalyst/resin ratio to be reduced as the mould warms up or the pigment colour to be rapidly changed

Higher volume machines use positive displacement pumps to measure individual quantities of resin, catalyst and pigment, which are then forced through a static mixer to blend the ingredients using compressed air and fed into the tool (*Figure 6.10*).

High volume production

For high volumes, a development of reaction injection moulding (RIM) *(Section 6.9)* has been introduced. This process makes use of modified RIM equipment in combination with a highly reactive resin system. The resin is heated and recirculated around delivery pipes to a high-pressure impingement mixer. A second circuit carries the curing agent which, after the two streams have met in the mixer, reacts with the resin and is cured in the heated mould.

For larger production runs, tooling is generally made in aluminium or by using nickel deposition on a master pattern. Heating and cooling coils can be built into the mould for the heated tooling needed to process fast-curing resin systems. Aluminium or steel tools manufactured by modern CAD/CAM equipment can also be produced, reducing the time required to move from concept to production.

Materials

Any thermosetting resin with a low viscosity can be used. Polyester resin is the most widely used for general applications together with epoxy, vinyl ester and phenolic for specific applications. The reinforcement can be of any type or form compatible with the resin system chosen.

Additives can be incorporated to enhance specific properties of the composite (*Table 6.4*). Materials such as chalk fillers can be used to improve both the surface finish and reduce costs. Fire retardants, pigments, and ultra-violet inhibitors can also be added.

Typical property data is given in *Tables 5.12* and *5.14*.

Summary: resin transfer moulding

Table **6.7**

Resin transfer moulding: advantages and disadvantages

advantages	disadvantages
closed mould	medium tooling cost
tight tolerances	high compatibility of fibre and resin
low emission of volatiles	medium production time
many types of reinforcement	
inserts possible	

6.5 Centrifugal moulding (CEM)

Centrifugal moulding or rotational moulding involves spraying a fibre/resin mixture onto the inner surface of a rotating mandrel (*Figure 6.11*). It is therefore an effective method of producing large cylindrical parts, with the added advantage that it produces two smooth surfaces from one tool face.

The process is ideally suited to the production of large diameter pipework and tank sections. It is also used to manufacture long tapered sections such as lamp posts *(Figure 6.12).*

Process

The process involves a rotating mould and a reciprocating boom which delivers the material and sprays it onto the mould surface *(Figure 6.1).* The nature of the process requires rugged tooling, and this tends to make it more expensive than the equivalent contact moulding tool.

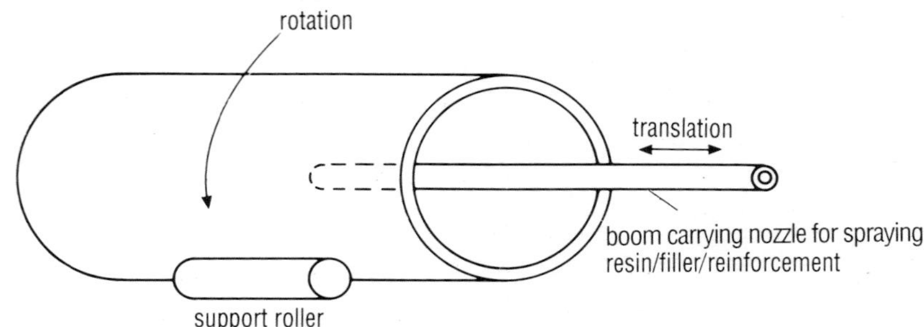

Figure **6.11**

Principle of centrifugal moulding process

The centrifugal forces produced while the materials are being laid down will generally be sufficient to wet out the reinforcement. However, additional compaction or rolling may be necessary if only slow rotation is possible. Rotation during cure also prevents the resin draining through gravity.

Materials

Most materials can be used. The most common ones are polyester resin, fillers and reinforcement, usually in the form of chopped glass fibres. It is possible to build up several layers with different properties in the component. This is useful in the chemical industry, where pipes may need a corrosion-resistant surface on both their outside surfaces and their internal bores. This is achieved by programming the boom to spray two gel coats, one on the outside and another, after the reinforcement has been built up, to act as the inside surface.

Another advantage of this process is that sand can be used as a cheap filler, sprayed in situ, to replace the more conventional chalk fillers which would be more difficult to apply unless premixed with the resin.

Figure **6.12**

Reinforced plastic lamp posts at Basingstoke (Lampro)

Summary: centrifugal moulding

Table **6.8**

Centrifugal moulding: advantages and disadvantages

advantages	disadvantages
large parts	cylindrical parts only
two smooth surfaces	must be rotatable
filler need not be pre-mixed with resin	limited range of reinforcement

6.6 Filament winding (FIW)

Filament winding involves the precise laying down of resin-impregnated reinforcement in the form of either rovings or tape onto a mandrel. It is a process for producing components with optimum properties using continuous fibre reinforcement. High structural efficiency can be obtained by orientating the reinforcement to match the direction and magnitude of applied loads.

Examples of filament-wound composite components include pressure vessels, drive shafts and pipes. While components do not necessarily have to be circular, these do tend to be the ones most widely manufactured. It is not possible to include undercuts, bosses and webs during winding, but it is feasible to incorporate inserts.

The component size range is enormous, varying from small-diameter tube to 40 m, 13 tonne wind turbine blades. The process produces a high quality component, but it requires a degree of automation for success. It is then a quick, effective and consistent way to impregnate and deposit large volumes of material.

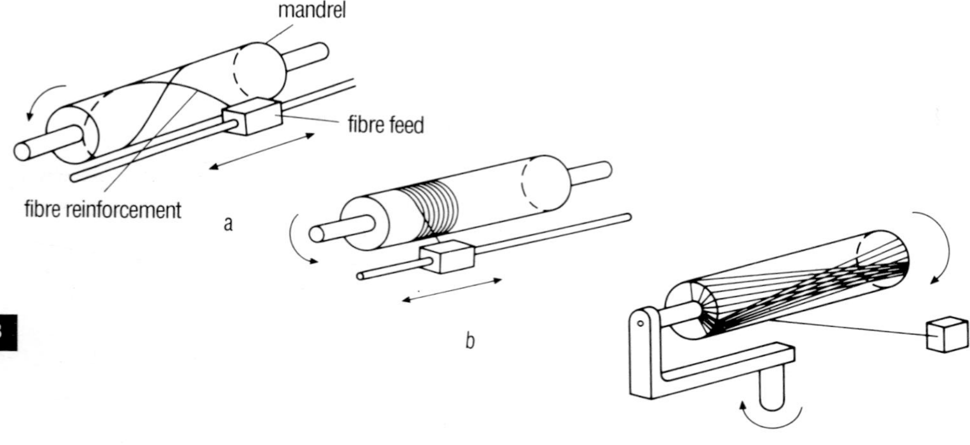

Figure **6.13**

Three principal types of filament winding pattern: (a) helical (b) hoop (c) polar (after Richardson)

Process

There are three main types of filament winding pattern – helical, hoop and polar – and these are shown schematically in *Figure 6.13*. The basic principles of each method are the same, the main difference being in the manner and direction of fibre lay-down. In its simplest form, a mandrel is rotated continuously in one direction while the fibre source reciprocates parallel to it. The rotation and reciprocation are synchronized to ensure that

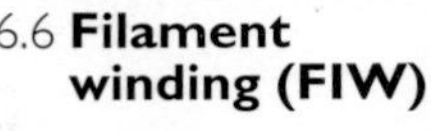

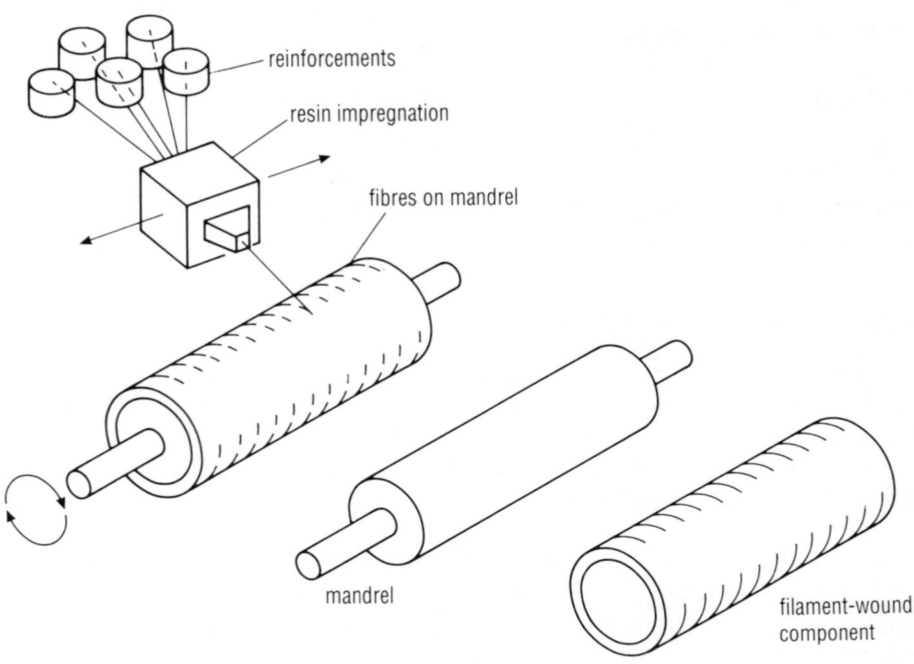

Figure **6.14**

A composite tube being helically wound over a removable mandrel (Richardson)

successive passes produce a fibre lay-down that gives progressive indexing to provide total cover of the mandrel surface. Additional freedom to wind more complex shapes can be obtained using a multi-axis winding spindle.

The fibre source may be a single roving, but it is more usually a group of rovings – up to several hundred are used for pipe winding. The process is continued until the desired material thickness has been achieved *(Figure 6.14)*. The mandrel may either be left in place after curing to form a permanent liner or, more usually, removed.

The normal method is to pull the component off the mandrel as soon as it has hardened. Techniques for removable mandrels include collapsible, segmented, and inflated designs, as well as ones employing melt out cores made of wax or low-melting-point alloys. In the case of thermosetting resins, the component is normally removed from the winding machine on the mandrel for post-curing.

Production equipment can be either simple or complex depending upon the type of winding and component being wound. The best results are obtained when the type of winding, the mandrel design and the winding equipment have been carefully matched.

The more advanced multi-axis winding machines are able to wind complex non-symmetrical shell structures under computer control.

Materials

Any continuous reinforcing fibre can be filament wound, and the process can be used to produce hybrid structures by winding several different fibres in turn.

Reinforcement can be applied in the form of tapes rather than rovings, as for example in winding a wind turbine spar *(Box 1.2)*. The two techniques are contrasted in *Table 6.9*.

Table 6.9
Winding with tape rather than rovings

advantages of tape winding	**disadvantages**
easier to handle	individual rovings cannot be tensioned
faster to lay down	lower property level
closer angles to winding axis	conformity to mandrel shape

The rovings are normally impregnated with a thermosetting resin before being wound. Polyester and epoxy are the two most common resins with others being used for more specialist applications. Fibre reinforced thermoplastic impregnated tape is now also available. With this material the resin needs to be simply softened prior to winding .

Property levels would be similar to those for the individual fibres *(Figure 5.1)*, corrected for fibre volume fraction and angle of winding *(Figure 3.8)*. *Table 5.14* gives some characteristic properties for carbon and aramid reinforced epoxy resins.

Summary: filament winding

Table 6.10
Filament winding: advantages and disadvantages

advantages	**disadvantages**
large lengths	cylindrical symmetry
large diameters	expensive tooling
accurate control of fibre emplacement	slow production rate without
complex geometry with multi-axis winding head	specialized facility

6.7 Pultrusion (PUL)

This is a method of manufacturing either discrete or continuous lengths of fibre reinforced plastics by pulling resin-impregnated fibres or fabrics through a series of forming dies. The final die is heated to cure the resin system, thereby producing a rigid composite section on leaving the die. The profile of the end product is determined by the tool cross section. The process is ideal for producing lengths of constant-profile sections, either solid or hollow, such as tubes, rods and structural sections. These can either be used immediately or fabricated into the required structure.

Examples of uses of pultrusions include aerial booms, skis, structural sections for buildings, aerofoils and ladders. It is possible to purchase lengths of standard sections and profiles from stock. Alternatively, profiles can be manufactured to a custom specification, although this involves the cost of the design and manufacture of the necessary tooling. A selection of typical profiles is given in *Figure 6.15 and Table 2.12*.

The minimum economic quantity of a custom profile depends on its size, performance requirements and the complexity of design. The maximum size of pultruded sections depends

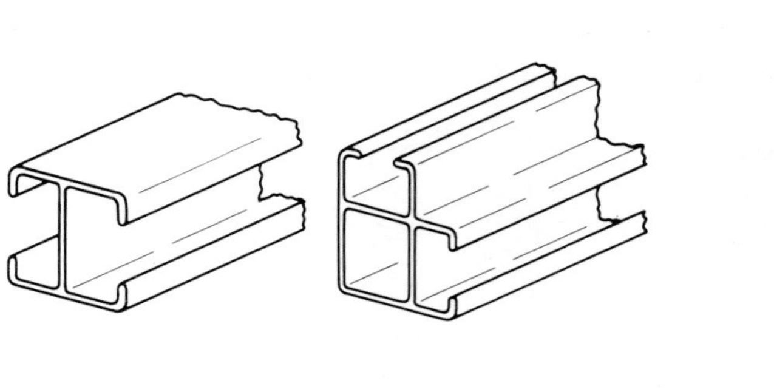

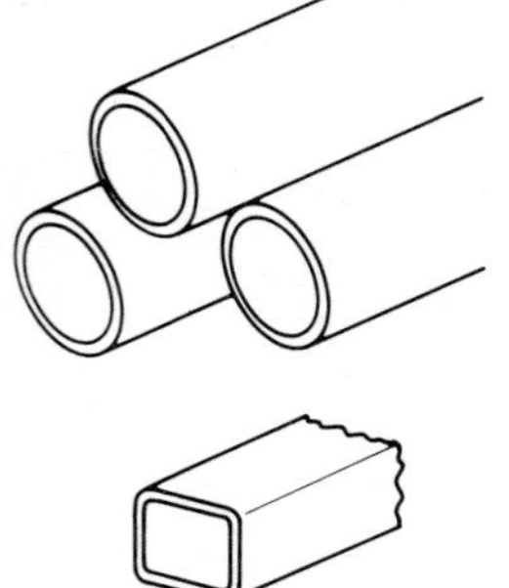

Figure 6.15

Typical pultruded profiles

on the capacity of the pullers, tool heating requirements, and number of roving ends (*Figure 3.2*). At present, commercial machines can produce pultrusions from 3 mm diameter rod up to hollow sections 1000 mm by 165 mm, with a pulling force up to 15 tonnes.

Process

The process starts by pulling continuous reinforcements, using either a belt or caterpillar puller system, through a resin bath to wet out the fibres. Rollers or pressure impregnators are used to ensure full wetting out of the fibre reinforcement.

Excess resin is then removed to expel any entrapped air and to compact the fibres. The impregnated reinforcement is then passed through preforming dies to align the reinforcement before it enters the heated curing section of the die.

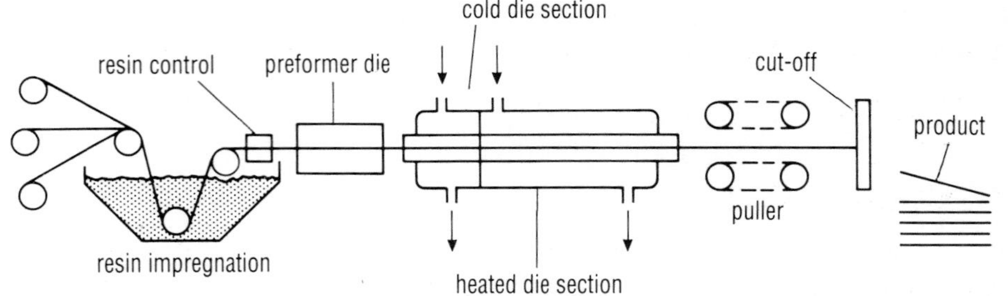

Figure 6.16

Pultrusion method in which reinforcement is impregnated by passing through a resin bath before entering the die (Richardson)

The temperature of the cure must be carefully controlled to prevent faults such as resin cracking, delamination, incomplete cure or sticking to the die surfaces from occurring.

Alternatively, the resin may be impregnated by pulling dry fibres into the preformer die and injecting the resin before the material enters the final heated die section.

The section produced can either be cut into discrete lengths after the puller system or wound onto a drum, as in the case of cable covering or small-diameter rods. The only limitation on length for cut sections is the space available after the cutters and for storage.

Materials

Composites with high stiffness and strength can be made using the pultrusion process as fibre volume fractions of up to 65% can be achieved with uni-directional aligned fibres.

Fibres such as glass, carbon or aramid can be processed in the form of continuous roving, woven cloth or mat, or any combination of these, to give a range of material and section properties.

Thermosetting resin systems such as polyester and epoxy are generally used in the pultrusion process. However, thermoplastics may also be used, with hot melt extrusion and pulling of the material.

Pulforming

Pulforming is closely related to pultrusion. It is a continuous process for the manufacture of curved composite components from predominantly axial aligned continuous reinforcing fibres. A major feature of this process is that the cross-sectional shape can be varied along the length of the component, although the cross-sectional area must remain constant.

The process was originally developed for the manufacture of composite automotive leaf springs. Hammer handles have also been made by this method *(Box 1.2).*

In a typical configuration, a number of female forming moulds are located on the circumference of a rotating wheel to form a complete die circle. The impregnated reinforcing fibres are pulled into the female mould as by a capstan through an orifice formed by the mould and a stationary die.

The combination of a rotatable mould and stationary die forms a closed cavity like the one in the pultrusion process. The composite material is cured while in the mould, normally using radio frequency curing, before being automatically cut to length and removed. When empty the moulds are recharged with resin-wet reinforcement and the process repeated.

The process is not widely used in the UK, and it has to be designed for each specific product, so it is normally only suitable for large-scale production. Components produced by this process have mechanical properties similar to those obtained from top quality pultrusions.

Availability of pultrusions

Many pultrusions are available from stock and the advantages of using such items has been discussed in some detail in *Section 2.4*. A set of typical sections is given in *Table 2.12* and typical properties in *Table 5.16*. For custom-built designs, the designer should seek early expert assistance from a manufacturer.

Summary: pultrusions

Table **6.11**

Pultrusions: advantages and disadvantages

advantages	disadvantages
variety of shapes	regular section only
readily available ex stock	difficult to join
high strength to weight ratio	tooling expensive for custom shapes
lightweight, easy to handle	
self coloured	

6.8 **Prepreg moulding (PPM)**

The consolidation of reinforcing fibre or fabric preimpregnated with resin (prepreg) can be achieved in two ways:

- by vacuum using a vacuum bag (vacuum bag moulding).
- by pressure in an autoclave (autoclave moulding).

The processes are commonly used for manufacturing components with a high degree of fibre alignment, because individual layers can be laid up in an optimal way *(Section 3.3)*. Higher pressures can be achieved with the autoclave than the vacuum bag, so autoclave forming is used for thicker samples or where a higher fibre volume fraction is required.

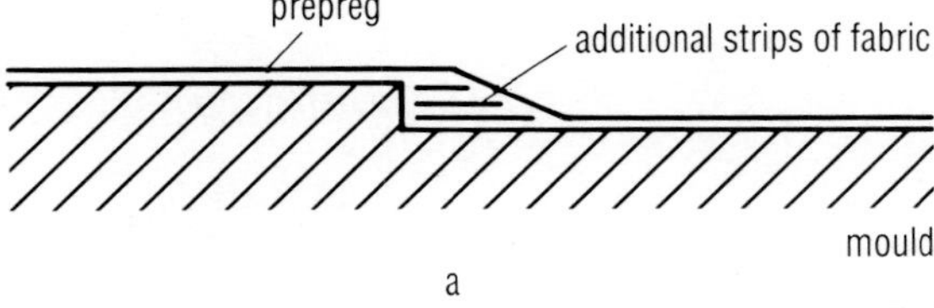

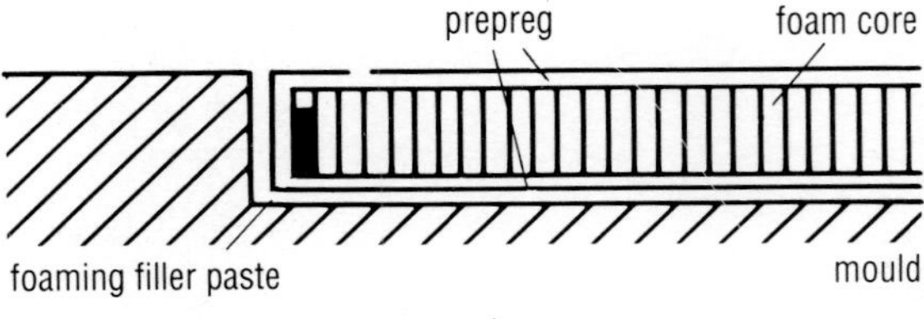

Figure 6.17

(a) Fabric tailoring method to infill corner
(b) Incorporation of foam core into a prepreg showing how core edge can be sealed (Noakes)

Both processes are extensively used in the aerospace industry where savings in mass are critical. The lower cost associated with the vacuum bag technique has enabled a similar quality to be achieved for more general applications such as high-speed cars.

Process

The process is simplified because the material manufacturer will have already determined the optimum mixture of resin, catalyst and reinforcement to achieve a specific range of properties when processed. Prepregs are in sheet or tape form with a certain level of tack (stickiness).

In its simplest form, the process is akin to laying up fabric in contact moulding *(Section 6.2)*. The prepreg sheets are first cut to a specific shape using a template. A gel coat layer is then applied to the mould surface and, as soon as this has partly cured, the layers can be laid on top of one another in the appropriate orientation. Where drape is insufficient, the prepreg has to be tailored *(Figure 6.17)*.

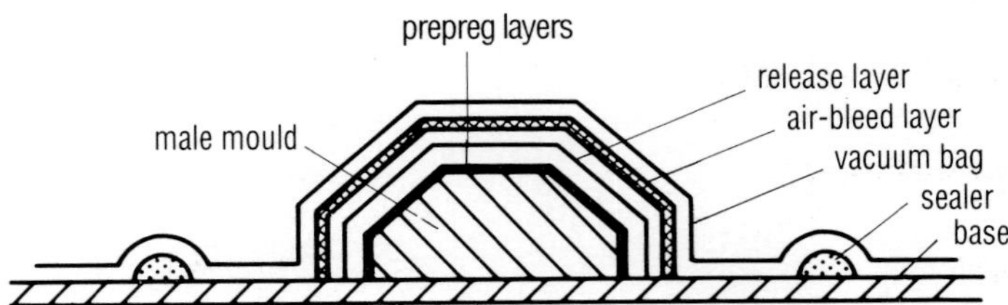

Figure 6.18

Method of forming vacuum bag showing how the prepreg layers are covered by a release layer and air bleed layer before sealing the bag onto a base plate (Noakes)

Consolidation may be achieved in one or more of the following ways:

- rolling with a ridged roller.
- pressing with a suitably shaped tool.
- vacuum bagging *(Figure 6.18)*.

Localized heating with a warm-air blower can be used to soften the resin if the level of tack is insufficient.
When the required thickness is built up, the bagged component is put in either an oven or an autoclave (with a pressure range up to 7 bar) and a controlled time/temperature cycle is initiated to cure the component.

The process can be semi-automated by using dedicated machines to cut the prepreg tape. Laying up can then be done by a pick and place robot.

Materials
A wide range of prepreg materials are available. Reinforcements include various types of fibres and fabrics, usually with epoxy resin, though many others have been used including phenolic (for low flame, smoke and toxicity) and polyimide (for high-temperature applications). Thermoplastic resins such as PEEK are starting to be used as well.

The lay-up sequence used allows the fibre orientation and type to be optimized, so prepregs are widely used with carbon and aramid as well as glass reinforcements and even in combinations to provide hybrid fabrics *(Section 5.2)*.

Sandwich panels are often produced with either honeycomb or foam cores to produce very light, stiff structures *(Figure 6.17)*.

Costs
The technique has a high labour content as the prepreg layers are thin and many layers are required to build up a reasonable thickness. The material cost is also high and prepregs have a shorter shelf-life than their constituent materials. While the vacuum bag method is relatively cheap, the cost of an autoclave is high.

In general the technique has only been used where low mass is at a premium and production runs are small.

Summary: prepreg moulding

Table **6.12**

Prepreg moulding: advantages and disadvantages

advantages	disadvantages
optimize fibre type	high materials cost
optimize fibre reinforcement	high labour content
low mass possible	many layers to place
optimize structure	high cost
reproducible properties	
established manufacturing route	

6.9 Injection moulding (IM)

This is a process in which a molten polymer and chopped fibres are injected into a mould where the composite is formed and allowed to cure. It is used for the high-volume production of small articles, often with an intricate shape, such as valves, pumps and fans.

Process

As with prepreg moulding, specially formulated moulding compounds (IMC) are used in which the appropriate constituents have been optimized *(Box 5.8)*. The basic process involves feeding the IMC granules into a heated barrel *(Figure 6.19)* which, together with the shearing action of the reciprocating screw, causes the mixture to melt. It is then injected into the mould.

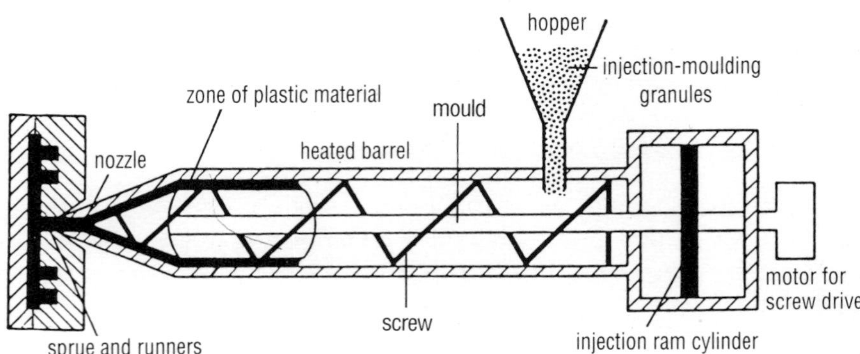

Figure 6.19 *Injection moulding machine showing the main parts (Philips)*

If a thermoplastic resin is used then the temperature of the mould is less than that of the melt, so the mixture begins to set as soon as it comes into contact with the walls of the mould. If a thermoset resin is used then the mould is heated to a temperature which triggers the chemical reaction and is then cooled to take heat out and set the moulding rapidly. Once the moulding is set, the mould is split and the part ejected.

Reinforced reaction injection moulding (RRIM)

In reaction injection moulding (RIM), two or more liquid intermediaries are combined in a low-volume mixing chamber where impingement mixing occurs. The materials are then fed at low pressure into a closed mould where they react and set, taking up every detail on the mould surface.

The introduction of reinforcement fillers such as milled or chopped glass into one of the polymer streams creates a reinforced reaction injection moulding (RRIM). By varying the formulation of the polymeric materials, a range of densities and stiffnesses can be obtained.

Compression injection moulding (CIM)

In this technique, a reciprocating screw is used to inject the molten compound into a mould that is held slightly open. Once the complete charge is injected, the mould is closed,

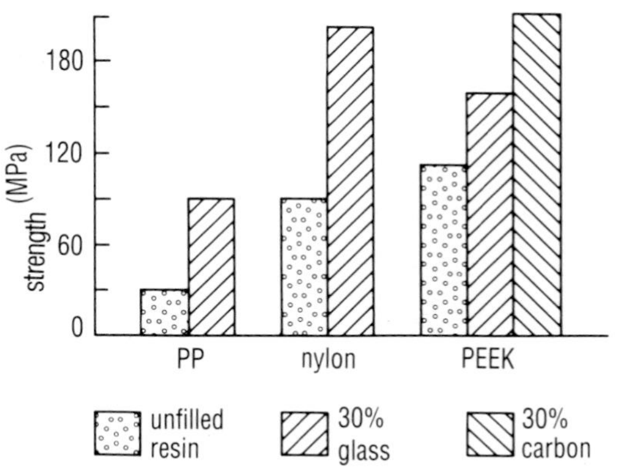

Figure **6.20**

Strength of some thermoplastic resins with either 30% (by weight) of glass fibre or carbon fibre. Flexural strength for plastic, tensile strength for reinforced plastic measured along direction of maximum fibre alignment (Booth)

forcing the melt into the cavities. By controlling the action of the screw and the way in which the mould is closed, the fibre degradation which tends to occur with injection moulding can be avoided and good mechanical properties can be obtained. The tailgates of the Citroen AX and BX *(Figure 1.1)* are produced in this way.

Materials

These processes all use injection moulding compounds, which are generally formulated by the resin supplier. Almost all resins can be used, but the most common are PA (nylon) and polypropylene (PP) amongst the engineering thermoplastics and PEEK amongst the advanced thermoplastics. There are so many grades of materials that the main manufacturers have developed databases to select the grade most suited for a specific need.

While thermosetting resins tend to have a slower manufacturing time, they have better heat resistance and are therefore used in applications where this is important. Polyester, in the form of dough moulding compound, is most widely used.

Fibres tend to be degraded by the mixing action, and fibre length tends to be reduced from about 10 mm to 0.25 mm, which is too short for efficient reinforcement. However, fibre length can be preserved by using special formulations in which each fibre is surrounded by resin, which lubricates its passage into the mould.

Typical fibre contents are approximately 30% by weight (20% by volume) and typical properties of some compounds are set out in *Table 5.15*. Carbon fibre has also been used as reinforcement and the strengths of some composites are compared with the basic strength of the resin in *Figure 6.20*.

Polyurethane is generally used with the RRIM process, often for automotive applications requiring energy absorption, such as bumpers.

Summary: injection moulding

Table **6.13**

Injection moulding: advantages and disadvantages

advantages	disadvantages
low labour cost	high capital cost
highly automated	restricted range of properties
high production volume	mould design very important
large variety of shapes	
self coloured	

6.10 Reference information

General references

Composites: a design guide, T Richardson, Industrial Press, New York, 1987.

Engineered Materials Handbook: Vol 2 Engineering Plastics, ASM International, Metals Park, 1987.

Hands Off GRP, Proceedings of an annual conference organized by PRI, London, 1989.

Design with Advanced Composite Materials, L Philips, Design Council, London, 1989.

Composites Manufacturing, Butterworth, London, 1990 – quarterly journal.

Specific references

6.1 Process selection

T Richardson *loc cit* – useful breakdown of product type and appropriate manufacturing process.

Design Manual for Engineered Composite Profiles, J Quinn, Fibreforce, Runcorn, 1988 – cost breakdown in terms of typical capital equipment and process.

6.2 Contact moulding

Polyester Handbook, Scott Bader, Woolaston, 1986 – detailed descriptions and lists good working practices.

T Richardson *loc cit* – comprehensive section on additives including selection table in terms of enhanced properties.

6.4 Resin transfer moulding

'Recent advances in glass fibre preforming', P Thornborough in *Hands Off GRP III,* PRI, London, 1989 – detailed description of process as applied to random mats.

'Preforming for resin injection moulding', D Jones in *Plastics and Rubber International,* August 1988 – good example of usage for general engineering products.

'RTM pumps up the volume', G Dean & R Heyman in *Advanced Composites Engineering,* Design Council, London, Summer 1988.

6.7 Pultrusion

J Quinn *loc cit* – good description of process, together with benefits and limitations.

6.8 Prepreg moulding

'Vacuum bag and autoclave moulding', N Brain in *IRPI,* Channel Publications, High Wycombe, 1985 – clarifies differences in processes and how to use them effectively.

Successful Composite Techniques, K Noakes, Osprey, London, 1989 – very comprehensive sections, full of practical considerations for manufacture.

6.9 **Injection moulding**

'Advanced composites – aerospace materials in industry', C G Booth & J S Kennedy in *Seminar S 721*, IMechE, London, 1989 – use of advanced thermoplastics in various manufacturing processes.

A description of the databases for short fibre reinforced resins is given in the reference section at the end of *Chapter 5*.

Testing and standards

chapter 7

summary

This chapter describes test methods and standards for reinforced plastics and composites, their component fibres and resin matrices. Testing at the sub-component, component and system levels are discussed, together with a brief review of non-destructive test methods. Published standards are listed in each section, with preference given to international standards.

Introduction

Previous chapters have identified the need for test methods and standards for various purposes. These include the initial design and material selection *(Chapter 2)*, detailed materials design data *(Chapter 5)*, development and approval of the production process and material, and assessment of the performance of the finished product or system. Ideally, a test method should satisfy more than one of the requirements in *Table 7.1*, as several will be required during the life of a project.

7.1 Standards and test methods

Table **7.1**

Test method: purpose

test requirements
- quality control and assurance
- material specification
- material development
- engineering design (design data)
- product development
- performance assessment
- research

Test methods may characterize the following:
- materials fabricated directly from fibres and matrices
- intermediate materials or compounds which are available as 'uncured' or 'un-consolidated' materials (such as DMC, SMC, GMT, and prepregs, as described in *Section 5.4*)
- finished products (for instance pultrusions, filament-wound tubes, sheets, and sandwich panels, as described in *Section 2.4*).

Depending on the chosen fabrication route *(Section 6.1)* characterization may be needed at several of the material levels detailed in *Table 7.2*.

Table **7.2**

Test method: materials

material types
- fibre
- matrix
- intermediate material (eg prepreg, SMC)
- laminate
- finished product (eg pultrusion, sandwich panel, filament-wound tube)

The major property areas are given in *Table 7.3*. It should be noted that properties other than stiffness and strength, such as shrinkage and thermal expansion, can also be anisotropic.

Most standards refer to glass fibres and thermoset matrices, because they have been available the longest. Until validated methods are available for the newer fibres and matrices, existing standards can normally be used in their original form or slightly modified provided that the required failure mode is obtained and any critical ratio, such as test span to specimen thickness, is maintained appropriately. However, this application is outside the scope of the standard. It may be necessary to measure some or all of the properties at the loading rate and temperature of the application, as well as under standard conditions *(Section 7.8)*.

Table **7.3**

Test method: properties

type	examples
mechanical	stiffness, strength, fatigue, impact resistance
thermal	expansion, conductivity
electrical	resistance, electrical strength
environmental	moisture and chemical resistances, temperature, ultra-violet resistance
process parameters	tack, flow, cure cycle
fire safety	smoke, emission, toxicity

Testing for engineering design purposes is generally undertaken on coupon specimens manufactured directly by injection moulding, or cut from either a test plaque or a moulded component. Separately manufactured specimens or test panels should be prepared using the same process as that planned for final production wherever possible, so parameters like the degree of compaction and voidage, and the cure cycle, are similar.

However, as the number of parts produced increases (see *Table 2.3*) from autoclaving through press moulding of moulding compounds to injection moulding of thermoplastics, the degree of flow occurring during manufacture increases, resulting in a greater likelihood of differences in fibre orientation and dispersion between samples and components.

7.2 **Material anisotropy**

As noted earlier *(Section 3.5)*, anisotropic materials need extensive data to characterize their properties accurately. For preliminary design purposes, testing can be minimized if the materials are approximately isotropic in-plane and, although layered, relatively thin *(Table 7.4)*.

In such cases, the ratio of the in-plane mechanical properties in perpendicular directions would be less than 2:1. However, it is possible to produce higher ratios of in-plane anisotropy through directional flow during processing, or by the intentional placement of additional aligned fibres.

Table **7.4**

'Isotropic' in-plane materials

'isotropic' in-plane materials
- injection moulded thermoplastics and thermosets
- sheet and dough moulding compounds (SMC and DMC)
- glass mat thermoplastics (GMT)
- mat and fabric (not unidirectional) thermosets

Table **7.5**

'Anisotropic' in-plane materials

'anisotropic' in-plane materials
- XMC, aligned GMT
- pultrusions (structural)
- unidirectional laminates
- multidirectional laminates
- filament windings

The highest ratios are obtained in unidirectional or multidirectional materials made from sheets of fully aligned prepregs *(Section 6.8)*, or made by filament winding *(Section6.6)* or pultrusion *(Section 6.7)*. For thin plate design in anisotropic materials, the minimum data for stiffness characterization are the Young's modulus in perpendicular directions, the in-plane shear modulus and an in-plane Poisson's ratio. These materials have also the highest ratio for in-plane to through-thickness (out-of-plane) properties *(Table 7.5)*.

The availability of suitable test methods is discussed in *Box 7.1*, and the current developments in standards for advanced polymer composites are given in *Box 7.2*.

Box **7.1**

Availability of test methods

Data requirements

Although test methods are available (though not yet necessarily as an international standard) for in-plane properties *(Section7.6)* for both isotropic and anisotropic materials, tests are not readily available for through-thickness tension, compression and shear properties. The availability of finite element analysis programs requiring comprehensive data inputs, and the use of reinforced plastics in thicker sections, are increasing the demand for through-thickness material properties, and associated test methods.

Anisotropy

For materials of low anisotropy (such as injection moulded thermoplastics, *Table 7.4*), test methods based on those for unreinforced plastics have been used, with caution, or have been adapted while noting the potential difference between moulded specimens and components.

For materials of high anisotropy *(Table 7.5)*, specialized test methods are often necessary to overcome specific problems. For example, when continuous fibre systems are tested along the fibre direction, end tabs *(Figure 7.3)* are normally used to overcome the difficulty of transferring the applied load into the specimens. In compression tests, additional care is needed to stabilize the specimen against gross buckling failures *(Figures 7.5 and 7.6)*. Weaker direction properties such as the transverse strength of aligned material may also need to be measured, even when good design has minimized off-axis loads.

Choice of test conditions

The anisotropic nature of reinforced plastics properties requires careful consideration of:

- the choice of test method for relevant/required materials data.
- the representation of service conditions.
- the design and conduct of the test itself.

These aspects are relevant to many properties, but they apply particularly to the range of stiffness and strength properties found in different directions within these materials. Extra care is required compared with testing isotropic metals or even polymers. The low stiffness and strength in certain directions (such as shear) makes it difficult to transfer the axial load into the specimen.

As minor secondary stresses can be critical in the lower-strength directions, it is important that tests of components or systems accurately reproduce the magnitude and direction of all stresses as far as possible. It is necessary to allow for the temperature *(Figure 7.17)* and rate dependence *(Figure 7.18)* which are particularly apparent for matrix-dominated properties. Even the tensile strength of GRP in the fibre direction is affected by loading rate: a tenfold increase in loading rate can increase the strength by approximately 10%.

7.3 **Pyramid of substantiation**

In all development programmes the test profile undertaken to develop and validate the component or structure will be pyramidal in shape *(Figure 7.1)*. Most testing will use laboratory coupon specimens, both in the initial design and development stage for material specification and engineering design data, and during the production period for quality control *(Sections 7.10 and 7.11)*.

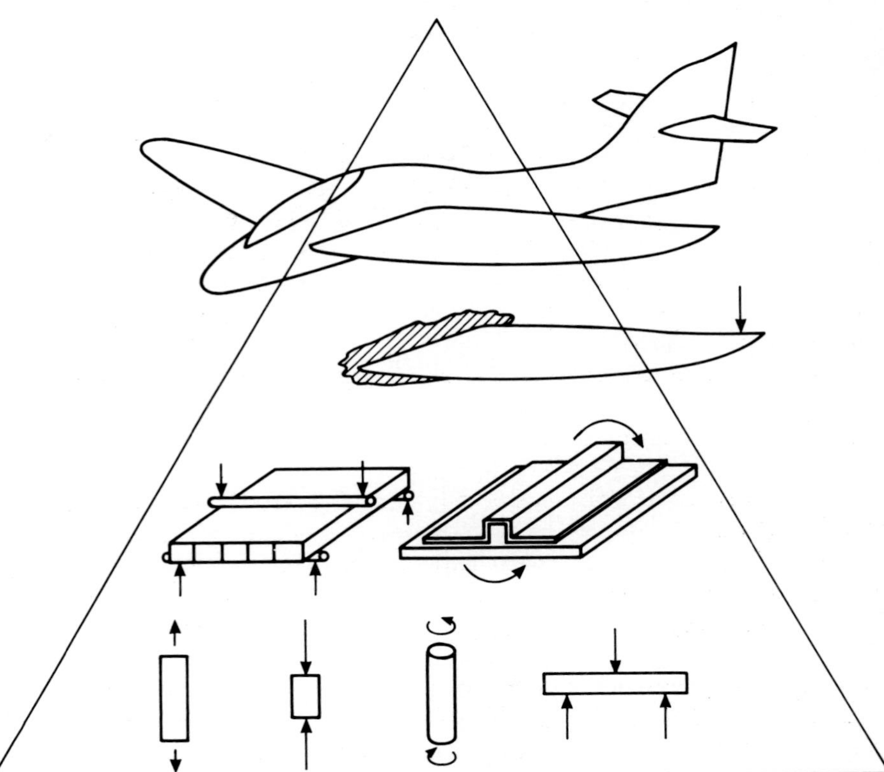

Figure **7.1**

Pyramid of substantiation in which smallest samples are tested the most and the largest structures the least

For large or complex structures, sub-component testing will be needed to validate specific design features *(Section 7.10)*. These sub-components generally represent the critical features of the design. Depending on their size and cost, the number of tests will be reduced compared with the amount of coupon testing undertaken. The highest level of validation is conducted on complete prototype or pre-production structures. Care must be taken that the loading accurately represents the actual service conditions as closely as possible.

The substantiation of the design must, if relevant, show evidence of the influence of long-term static loads, repeated (fatigue) loads, impact damage and environmental attack, for example by chemicals, moisture, or ultra-violet radiation *(Sections 3.5 to 3.9)*. The interaction of these effects must also be checked if two or more cause degradation of the material properties.

Care must be used when attempting to accelerate the effect of any of these processes – for example by using higher fatigue stresses or test frequencies *(Box 4.6)* – that no change in the failure mechanism occurs *(Section 3.4)*. This is particularly important when more than one effect is being studied simultaneously.

Box **7.2**

Development of material test standards

Applicability

The standards with the widest application are the world-wide ISO (International Standards Organization) series and the CEN (Comité Européen de Normalisation) series for use in CEC and EFTA countries. The ISO TC61 series is fairly comprehensive for GRP, but ISO Technical Committee only began in 1990 to consider test methods for carbon fibre reinforced plastics (CFRP) and other advanced composites.

The ISO standard-making process is rather slow because there is only a single annual meeting to confirm new standards for publication. The EN series produced by CEN is likely to be quicker, particularly in the Reinforced plastics – Aerospace series, due to an active trade group (AECMA) and the need to produce standards in support of the Single European Market *(Section 4.4)*. A CEN committee for the general engineering use of composites, TC249/SC2 has been established. This committee has established a work programme based on harmonization with ISO test methods wherever possible.

Advanced composites

In both international series (ISO, CEN) the delegates and voting rights reside in the national bodies. The most active with their own national series are AFNOR, ASTM, BSI, DIN and JIS *(Box 4.3)*.

In addition there are other groups actively promoting recommended test methods, such as:

CRAG (Ministry of Defence – Composites Research Advisory Group) (UK)
SACMA (Suppliers of Advanced Composite Materials Association) (USA)
ACOTEG (Advanced Composites Technology Group) (European)
AECMA (Association Européenne des Constructeurs de Material Aerospatial) (European)

These groups are concerned with advanced composites. They each aim to persuade a wider cross-section of organizations to use their methods and to turn them into formal standards through the national and international standard bodies.

There are also many company specifications and test methods in use. However, as many of these methods are related to, or have been derived from, other test methods, there is now a major attempt to harmonize methods internationally. For example, ASTM test methods are already in wide international use, as are those from the Boeing Company and Airbus Industrie.

Validation
The increased importance of standards within the legal framework for free trade agreements and/or product liability has resulted in greater adherence to the requirement for test methods to be validated by round-robin tests. The precision of the method so established is stated and has to be commensurate with its intended use (ISO). Harmonized and more up-to-date standards should come into being through the international pre-standards research being under taken through the VAMAS initiative (for example on fatigue test methods).

7.4 Fibre, matrix and intermediate materials test methods

Fibre test methods

These test methods are normally used by the material supplier, the fabricator, the researcher, or as part of the quality control/material specification/process control stage of the project, rather than by the designer. However, knowledge of the material properties of the constituent phases can be used to predict the expected properties of the reinforced material *(Box 3.2)*. This is well established for continuous fibre systems, where several laminate calculation programmes using desktop computer software are available *(Section 3.10)*.

Test methods for glass fibres and their associated yarns, rovings and cloths are well established, both nationally and internationally. Similar standards are being developed for carbon fibres and aramid fibres, but new standards are required for their unique properties, such as the thermal oxidation resistance of carbon fibres. The main fibre properties are given in Table 7.6.

Table **7.6**
Main fibre properties

main fibre properties
- density or mass per unit length
- diameter and shape
- axial strength
- axial stiffness
- size content

The mechanical properties can be measured for individual fibres or for a strand or roving. The latter is the easier strength test to conduct, but it requires a strand or roving to be impregnated with resin and cured. It should be noted that, as the measured strength of these fibres decreases with increasing gauge length, a length similar to the equivalent laminate coupon specimens should normally be used. Properties which characterize fibres and mats, yarns and fabrics, are listed in *Tables 7.7* and *7.8*.

Box **7.3**
Tabulation of standards

Applicability of standards
Published standards are listed in *Tables 7.7* to *7.20*, with preference given to international standards where these exist because of their their increasing importance for free trade and product liability purposes. EN standards, for example, have to be published as national standards by member countries, and any existing national ▶

standard of the same scope withdrawn. The standards are drawn from several series, so a code has been used to show the application area in all the following tables:

C = carbon fibre or carbon fibre reinforced plastics
G = glass fibre or glass fibre reinforced plastics
F = all fibres or all fibre reinforced plastics
P = plastics
RP = reinforced plastics
* = *draft versions or study items*

Titles
The titles of standards are often long in order to state their purpose and scope clearly. They have been shortened here to give only essential information. Most ISO standards are listed under Field No 170 'Plastics'. The current draft European standards listed (EN*) are in the Aerospace series. Where no appropriate standard has yet been produced, careful use of existing methods is possible while ensuring that test criteria and required failure modes are met, although such use will be outside the standard.

While standards in ISO Plastics series often apply to both unreinforced and reinforced plastics, reinforced plastics are sometimes mentioned explicitly in a standard.

Table **7.7**
Fibre standards

title	number	scope
definitions and vocabulary	ISO*	(C)
fillers and reinforcing materials – symbols	ISO 1043/2	(P)/(RP)
batch sampling	ISO 1886	(G)
combustible matter content	ISO 1887	(G)
	EN* 2967	(C)
size content	ISO* 10548	(C)
average diameter by cross-section	ISO 1888	(G)
	EN* 2965	(C)
diameter and shape	ISO* 11567	(C)
density of high modulus fibres	ASTM D3800	(F)
E-glass fibres – specifications	BS 3496	(G)
density	ISO 10119	(C)
linear density (mass per unit length)	ISO 1889	(G)
	EN* 2964	(C)
	ISO 10120	(C)
textile glass-moisture content	ISO 3344	(G)
rovings – basis for specification	ISO 2797	(G)
rovings for polyester and epoxy resins	BS 3691	(G)
tensile strength of impregnated yarn	EN*	(C)
tensile strength of impregnated rovings	ISO*	(G)
thermal oxidation resistance	ASTM D4102	(C)
tensile properties-filaments, strands etc	ASTM D4018	(C)
reporting test methods and results	ASTM D3544	(F)
strength and modulus of filament	ASTM D3379	(F)
	ISO* 11566	(C)

Table **7.18**

Mat, yarn and fabric standards

title	number	scope
mats– dissolution of binder in styrene	ISO 2558	(G)
– basis for specification	ISO 2559	(G)
– tensile loading force	ISO 3342	(G)
– mass per unit area	ISO 3374	(G)
– thickness and recovery	ISO 3616	(G)
mats and fabrics – contact mouldability	ISO 4900	(G)
woven fabrics – yarns per unit length	ISO 4602	(G)
– thickness	ISO 4603	(G)
– flexural stiffness	ISO 4604	(G)
– mass per unit area	ISO 4605	(G)
– tensile properties	ISO 4606	(G)
– width and length	ISO 5025	(G)
– basis for specifications	ISO 2113	(G)
rovings – specifications	ISO 2797	(G)
– stiffness	ISO 3375	(G)
twist of yarns	ISO 1890	(G)
yarns – designations	ISO 2078	(G)
– breaking force and elongation	ISO 3341	(G)
– twist balance index	ISO 3343	(G)
– basis for specification	ISO 3598	(G)
textured yarns – basis for specification	ISO 8516	(G)
tensile properties of impregnated yarns	ISO*10618	(C)

Resin matrix test methods

The main properties for the use of resins as matrices are given in Table 7.9.

Table **7.9**

Main matrix properties

- density
- viscosity
- volatile content
- heat distortion temperature
- chemical resistance (if required)
- modulus
- strength
- failure strain

Many specifications and test methods exist for polymer matrices, both in the general plastics series and in the individual series for a particular classification (such as polyesters). These standards are not referred to here because they are listed in detail in the standard indexes and because the range of materials used as matrices is continually increasing.

Table **7.10**

Resin matrix standards

title	number	scope
water absorption	ISO 62	(P)
deflection under load temperature	ISO 75	(P)
flexural properties	ISO 178	(P)
tensile properties	ISO 527	(P)
compression properties	ISO 604	(P)
density and relative density	ISO 1183	(P)
polyester resins – designations	ISO 3672	(P)
epoxide resins – designations	ISO 3673	(P)
fracture toughness	ISO*	(P)
differential scanning calorimetry	ISO*	
mould shrinkage	ISO 2517	(P)

The most relevant matrix standards for composite applications are listed in *Table 7.10*. Standards are being developed for the differential scanning calorimetry technique, which can be used to check correct cure of a thermoset system *(Table 5.7)* and the age of a prepreg *(Table 7.11)*.

Interfacial properties

No standard test methods are available for directly measuring the interfacial properties between fibre and matrix. These properties are important as they affect the reinforcement efficiency and the relative toughness (crack growth resistance) of the final material. Some indirect tests such as short-beam interlaminar shear, flexural strength and delamination fracture toughness *(Table 7.14)* are used to assess the interface properties.

Intermediate materials and compounds

Several properties have been defined for thermoset systems relevant to either their specification or processability. These properties are given in *Table 7.11*.

Table **7.11**

Intermediate material properties

property	definition
gel time	point at which resin starts to harden
tack level	stickiness of prepreg
resin flow	flow of resin in prepreg
cure cycle	time/temperature profile for curing
volatile/resin/fibre contents	prepreg composition
glass transition temperature	transition temperature for glassy to rubbery behaviour of polymers

The majority of test methods for intermediate materials concern either the mass and mass fractions of the prepreg or relate to its processability. Relevant standards are given in *Table 7.12*.

title	number	scope
prepregs – definitions	ISO 8604	(P)/(RP)
– mass per unit area	EN* 2329	(G)
– mass per unit area	EN* 2557	(C)
– volatile matter	EN* 2330	(G)
– volatile matter	EN* 2558	(C)
– volatile matter	ASTM D3530	(C)
– resin content	EN* 2331	(G)
– resin content	ASTM D3529	(C)
– resin content	ASTM C613	(C)
– resin flow	EN* 2332	(G)
– resin flow	EN* 2560	(C)
– resin flow	ASTM D3531	(C)
– resin/fibre per unit area	EN* 2559	(C)
– gel time	ASTM D3532	(C)
SMC – specification basis	ISO* 8605	(G)
DMC – specification basis	ISP* 8606	(P)

Table **7.12**

Intermediates and compounds standards

7.5 Coupon test methods for mechanical properties

A wide range of coupon test methods is available. Some, for reinforced plastics have been based on, and are often included within, general plastics test methods, in particular for the better established GRP materials. As noted in *Box 7.2*, there is strong interest in a specialized series of standard test methods for advanced polymer matrix composites.

Preparation of test panels and specimens are covered in *Box 7.4*, and the major standards for test methods for cured and/or consolidated laminates in *Table 7.14*.

Box **7.4**

Specimen preparation

Test panels
If the coupon is not taken from an actual product nor moulded to size, it will be necessary to prepare a test panel. Some standards are available for different processing routes *(Table 7.13)*, but versions are not yet published for non-glass fibre systems.

Preparation
No specific standards are available on specimen preparation, although machining of plastics is covered by ISO 2818.

Diamond slitting and grinding wheels, normally with water lubrication, have been found to be satisfactory, and water jets and laser cutting have also been used successfully. Special techniques are needed for aramid fibre composites, and information is available from the material suppliers.

It is very important, particularly when dealing with high-performance materials, that the test panel is accurately made and that the specimen is accurately cut along a known ▶

reference direction (normally the 0° direction or, for a continuously produced sheet, along the length of the sheet). For fully aligned material it may be necessary to trace the fibre tows or break off a panel edge to determine the 0° fibre direction.

Table **7.13**

Panel and specimen preparation standards

title	number	scope
compression moulding – thermoplastics	ISO 293	(P)
compression moulding – thermosets	ISO 295	(P)
injection moulding thermoplastics	ISO 294	(P)
low-pressure laminated plates	ISO 1268	(G)
press moulded/autoclave panels	EN* 2565	(C)
	EN* 2374	(G)
preparation of specimens by machining	ISO 2818	(P)
plates by filament winding	ISO*	(G)
unidirectionally reinforced plates	ISO*	(G)
ring test specimen	ASTM D2291	(G)

Note: ISO 1268 is being revised in 9 parts to cover most methods for all materials.

Testing

Care is also required when attaching tabs or strain gauges. It has been reported that erroneous results can be obtained using strain gauges, but they are still in extensive use. Clip-on extensiometers are used instead of strain gauges and lead to large cost savings.

In all cases specimens should be tested in two perpendicular directions if the material is considered to be anisotropic in the plane; one of these directions should be the principal fibre direction. This cannot be done with injection-moulded beam or dumb-bell specimens, where the actual moulding gives preferential alignment along the specimen resulting in relatively high values. Properties more typical of moulded components should then be obtained from samples cut from moulded plates (150 × 150 mm or larger).

Table **7.14**

Coupon test standards

title	number	scope
flexural properties	ISO 178	(P)/(RP)
	EN* 2562	(C)
	EN*2746	(G)
impact strength – Charpy	ISO 179	(P)/(RP)
impact strength – Izod	ISO 180	(P)/(RP)
multi-axial impact, instrumented	ISO 6603	(P)
tensile properties	ISO 527	(P)
	ISO 3268	(G)
	EN* 2561	(C)
(transverse)	EN* 2597	(C)
	ASTM D3039	(F)
(ring)	ASTM D2290	(G)
torsion properties	ISO 537	(P)
compressive properties	ISO 604	(P)

compressive properties	EN* 2850	(C)
	ISO 8515	(G)
	ASTM D695	(F)
	ASTM D3410	(F)
compression after impact	DIN 65561	(F)
open-hole tensile strength	DIN* 65559	(F)
closed-hole compressive strength	DIN* 65560	(F)
deflection temperature	ISO 75-3	(RP)
interlaminar shear	ISO 4585	(F)
	EN 2377	(G)
	EN* 2563	(C)
	ASTM D2344	(F)
in-plane shear	ASTM D3518	(F)
	ASTM D4255	(F)
fibre content – by digestion	ASTM D3171	(F)
– by electrical resistivity	ASTM D3355	
– by loss on ignition	ISO 1172	(G)
void content	ISO 7822	(G)
density	EN* 2745	(RP)
moisture content	EN* 3616	(RP)

Note: ISO 527 is being revised to include ISO 3268 as part 4

Figure **7.2**

Untabbed specimen design (based on ISO 3268). (a) dumb-bell (Type 1); (b) plain (Type 2); (c) cross section

Individual test methods

This section briefly reviews the main aspects of various coupon test methods.

Tensile tests (eg ISO 3268, ISO 527, ASTM 3039):

- provide basic engineering property data for use in design or in material specifications.
- dumb-bell type specimens *(Figure 7.2)* are suitable for isotropic in-plane materials *(Table 7.4)*. Plain strip specimens can also be used provided that failure at the grip is avoided *(Figure 7.2)*.
- for unidirectional and multidirectional materials it is normally necessary to use parallel-sided specimens with end tabs *(Figure 7.3)*. The through-thickness waisted specimen for 0° material is no longer favoured as the measured values are higher than those achieved in the 0° ply in multidirectional specimens or components.
- grip pressure must be limited to avoid crushing of aligned specimens due to their low transverse strength. Through-bolted connections are unsuitable for most materials.
- alignment of specimens is important, particularly in the transverse direction of unidirectional material, due to the low failure strain and the large width of the specimens.

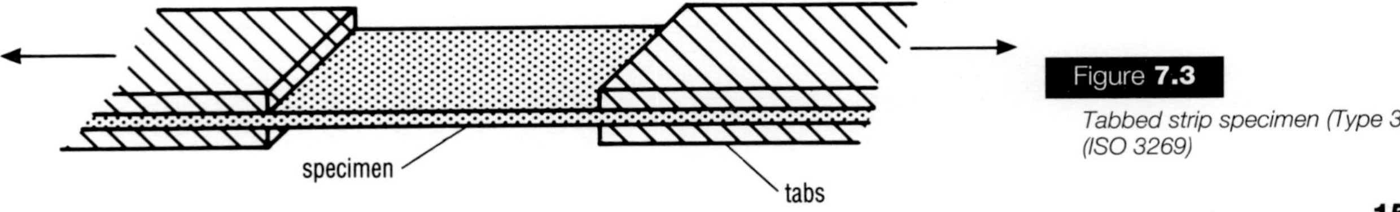

Figure **7.3**

Tabbed strip specimen (Type 3) (ISO 3269)

Figure **7.4**

Alternative compression failure modes

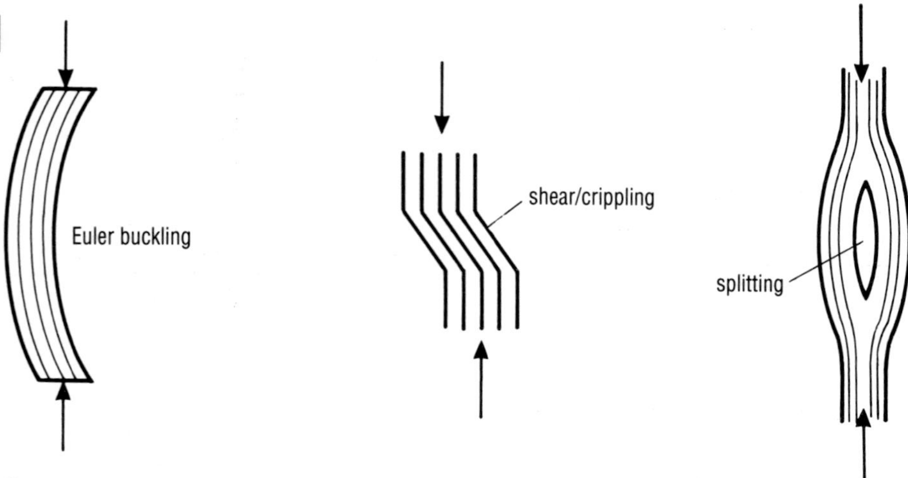

Compressive tests (eg ASTM D3410):

- provide engineering property data.
- specimens tested at short gauge lengths as long unsupported specimens are susceptible to Euler buckling *(Figure 7.4)*.
- several support or anti-buckling jigs are available *(Figures 7.5 and 7.6)*. Celanese or IITRI type jigs (ASTM D3410) are normally preferred.
- unidirectional materials tested at short gauge lengths fail by shear/local crippling (kink-band) failures *(Figure 7.4)*.
- no standard test method is available for splitting (delamination) failure *(Figure 7.4)*.
- initial specimen quality is important and strain gauges on both sides of the specimen are used to monitor eccentricity of loading *(Figure 7.7)*.
- loading can be by shear (ASTM D3401), end loaded (ISO 8515) or a combination.

Figures **7.5 & 7.6**

Anti-buckling jig (based on ASTM D695 – modified)

IITRI anti-buckling jig (ASTM D3410)

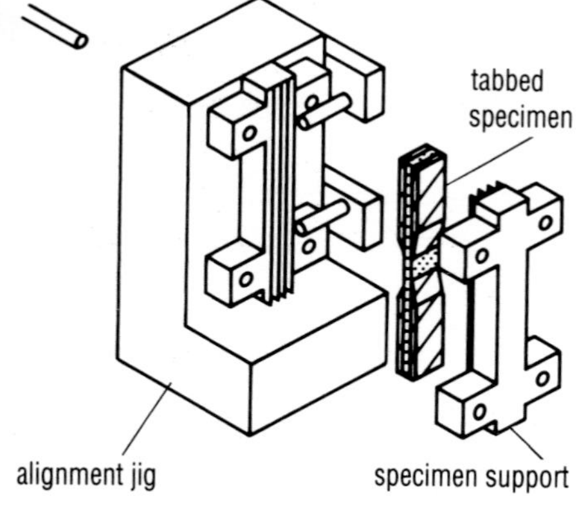

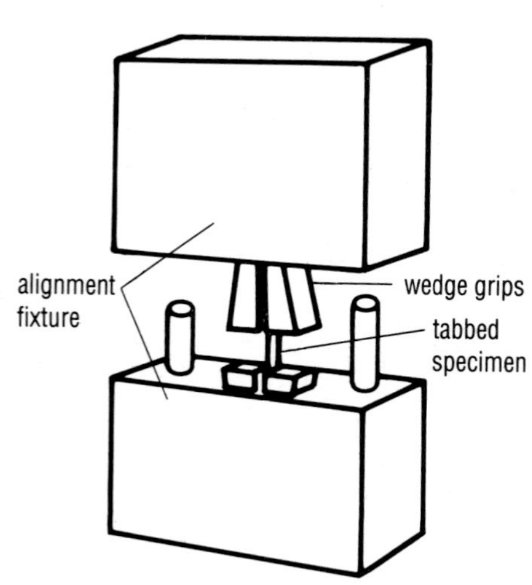

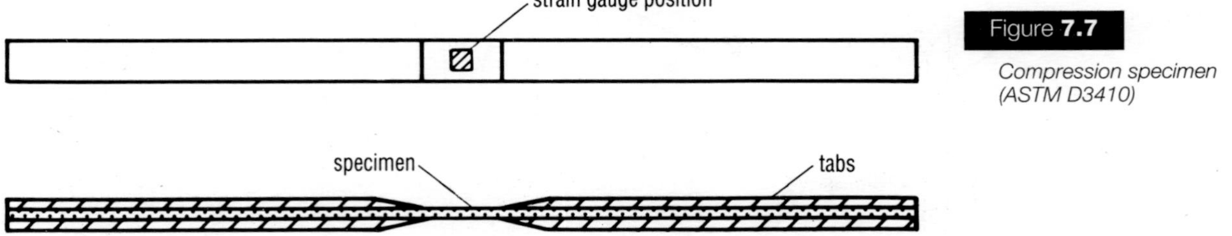

Figure **7.7**

Compression specimen (ASTM D3410)

Flexural tests (eg ISO 178)

- normally used for quality control testing rather than engineering property data.
- normally conducted under three-point loading (*Figure 7.8*). Four-point loading reduces the likelihood of compression face failure (mostly in fabrics or 0° aligned systems) and gives pure bending between the centre loading points.
- minimum span-to-depth (l/d) ratio is necessary for accurate measurement of flexural modulus so that deflection due to shear is minimized (eg l/d≈40/1 for CFRP and 20/1 for GRP).
- possible to test large coupons (or suitably shaped components).
- sample preparation is easy.
- SACMA *(Box 7.2)* has declined to publish a method because of the large effect of test conditions on recorded values.

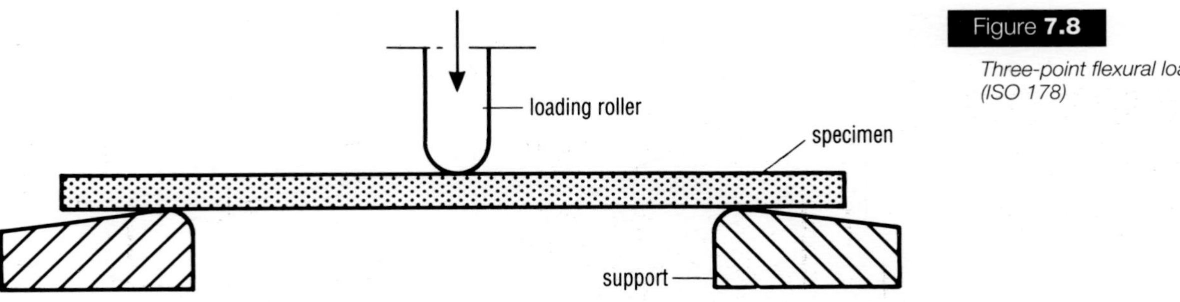

Figure **7.8**

Three-point flexural loading (ISO 178)

Shear tests (eg ASTM D3518)

- wide range of methods available.
- ±45° tension, thin walled cylinder in torsion or plate twist preferred for in-plane properties *(Figure 7.9)*.
- Iosipescu test suitable for laminated and non-laminated materials.
- most direct block, scissor, or bend shear tests *(Figure 7.9)* give average values and do not allow for the actual stress distribution.
- selected test method should be appropriate to the shear property required (i.e. in-plane or through-thickness.)

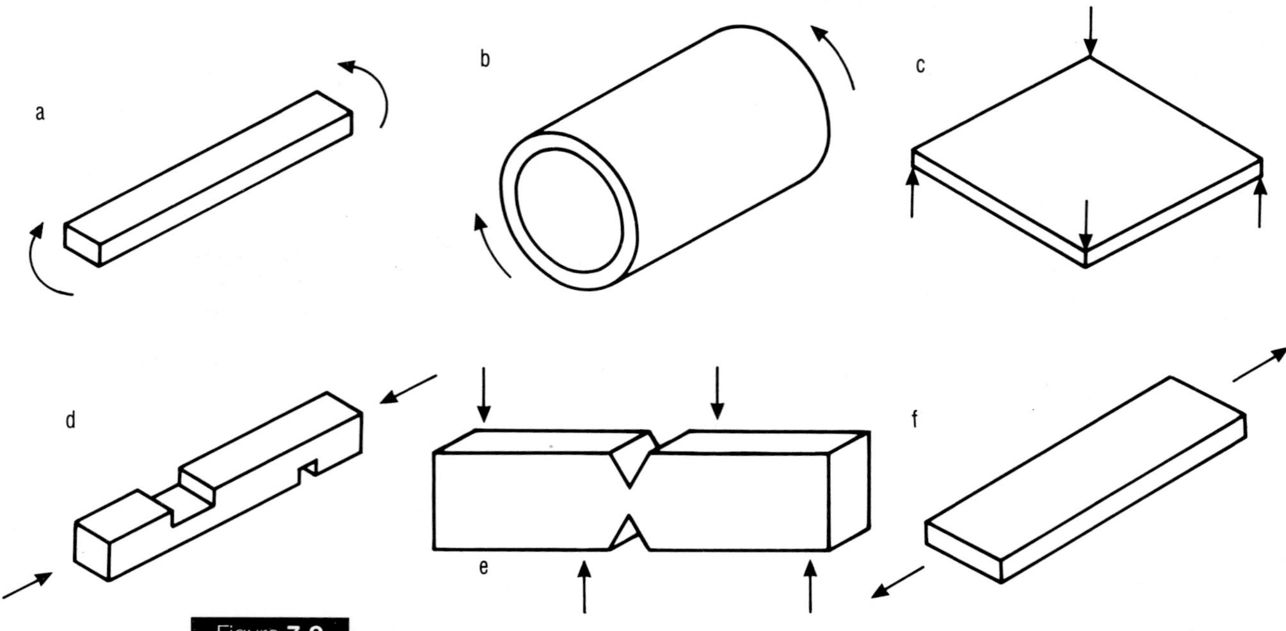

Figure 7.9

Selection of shear test methods. (a) beam torsion; (b) thin walled tube torsion; (c) plate twist; (d) scissor shear; (e) Iosipescu/asymmetrical bend; (f) ±45°/10° off-axis tension

Interlaminar shear strength test (ILSS) (eg ISO 4585, EN* 2377)

- normally used for quality control testing rather than for engineering data.
- several failure modes are possible, including tension, compression or combined failure modes, but only shear failures are acceptable *(Figure 7.10)*.
- low span-to-depth ratio (5 or 4) used to encourage interlaminar shear.
- shear modulus is not measured, and calculated shear stress assumes the stress distribution applicable to an isotropic material specimen.

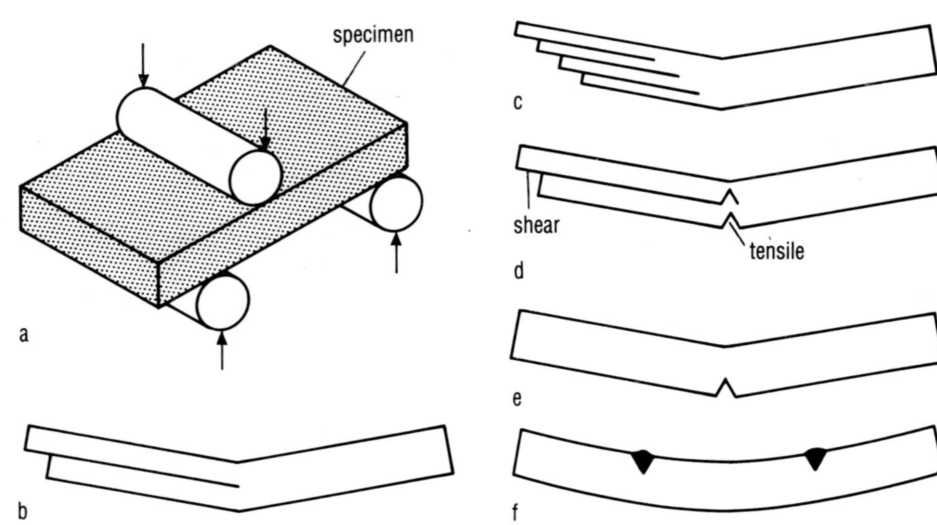

Figure 7.10

Interlaminar shear test standard (based on EN 2377). (a) Interlaminar shear test. Valid failure modes: (b) single shear; (c) multiple shear; (d) shear and tensile. Invalid failure modes: (e) tensile crack; (f) compressive cracks.*

Impact tests (eg ISO 179, ISO 180, ISO 6603)

- Izod (ISO 179) and Charpy (ISO 180) *(Figure 7.11)* are suitable for isotropic and laminated materials *(Table 7.4)* tested in-plane, but are not suitable for unidirectional material or out-of-plane measurements of laminated materials as notches are not always effective.
- instrumented falling weight impact test (dart) now preferred, but recommended span (40 mm diameter in ISO 6603) is possibly too narrow. Test used for a wide range of materials with 60 mm diameter or 60 x 60 mm plaques (also referred to as a puncture test).

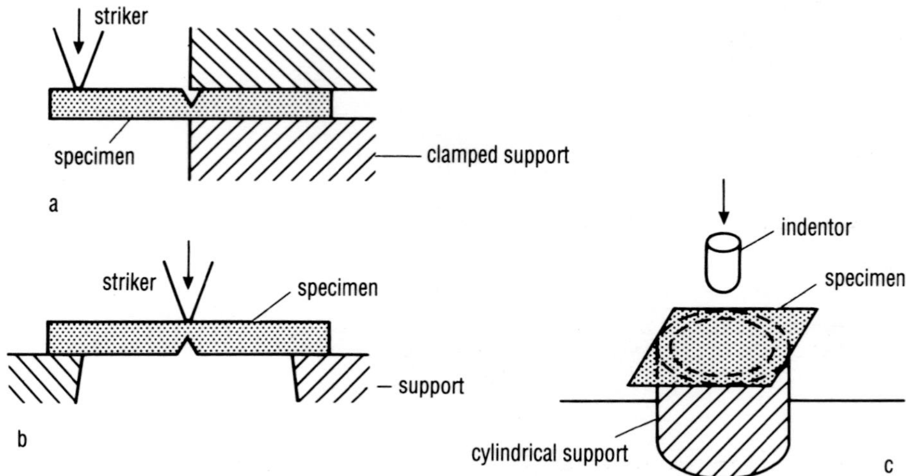

Figure 7.11

Impact test methods: (a) Izod (ISO 179); (b) Charpy (ISO 180); (c) puncture (ISO 6603)

Compression after impact test (eg DIN* 65561)

- devised for aerospace applications *(Figure 7.12)* in which the specimen is first impacted with an indenter and subsequently tested in compression using an anti-buckling support.
- no mechanical property data taken from impact phase of test, but area and type of damage noted.
- residual compression properties measured.
- Both ASTM and DIN are preparing standards based on the Boeing method (1992).

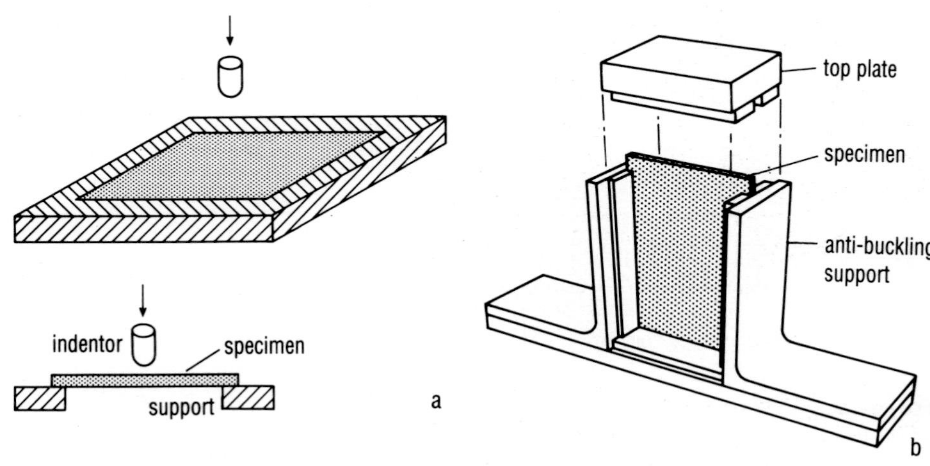

Figure 7.12

Compression after impact: (a) impact phase; (b) compression phase (DIN 65560)

Tensile or compressive holed specimen test (eg *DIN 65559/DIN* 65560)

- provides engineering data on allowable strain limits on specimens which have a hole drilled through them *(Figure 7.13)*.
- uses a multidirectionally reinforced specimen, with 'open' or 'filled' hole.
- developed in aerospace industry.
- compressive-loaded specimens must be supported against buckling.

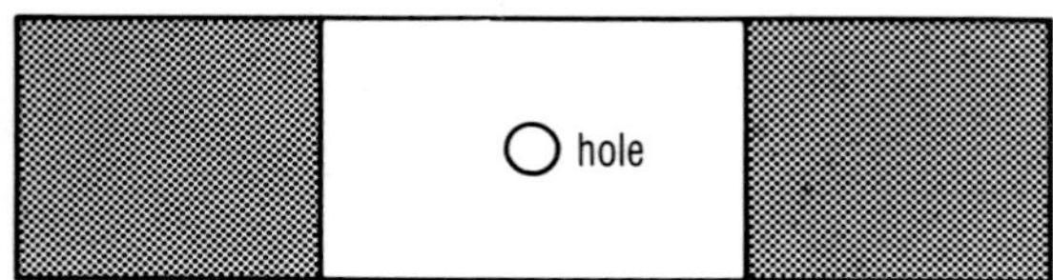

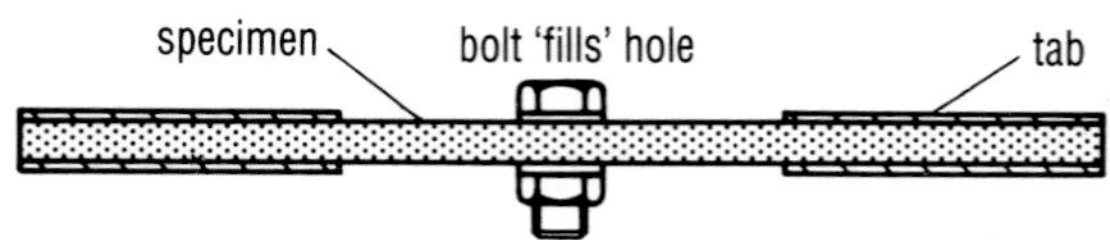

Figure **7.13**

Multidirectional compressive specimen with filled hole (DIN 65561)

Bearing test (eg CRAG 700)

- refers to the testing of bolted connections in which a bearing is inserted into a specimen and a bolt is then added and torqued *(Figure 7.14)*. Failure is defined as excessive deflection of the bolt hole under a remotely applied tensile load.
- uses a multidirectionally reinforced laminate specimen.

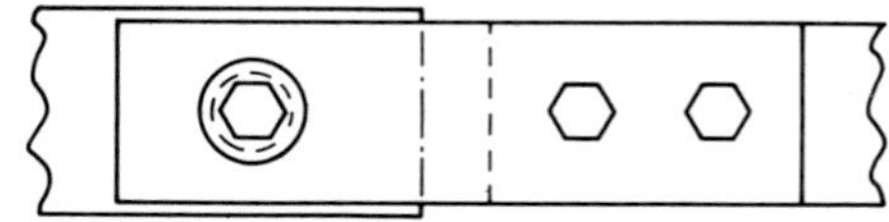

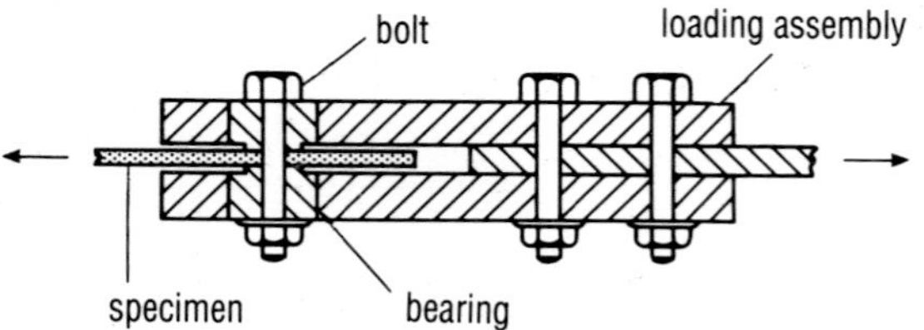

Figure **7.14**

Bearing test showing bolt through specimen (CRAG 700)

Fracture toughness tests (eg ISO*)

- test measures energy required to produce failure in a pre-notched sample *(Figure 7.15)*.
- standard methods in draft form (1990) for polymers and possibly isotropic composites (*Table 7.4*); geometry similar to metal specimens *(Figure 7.15)*.

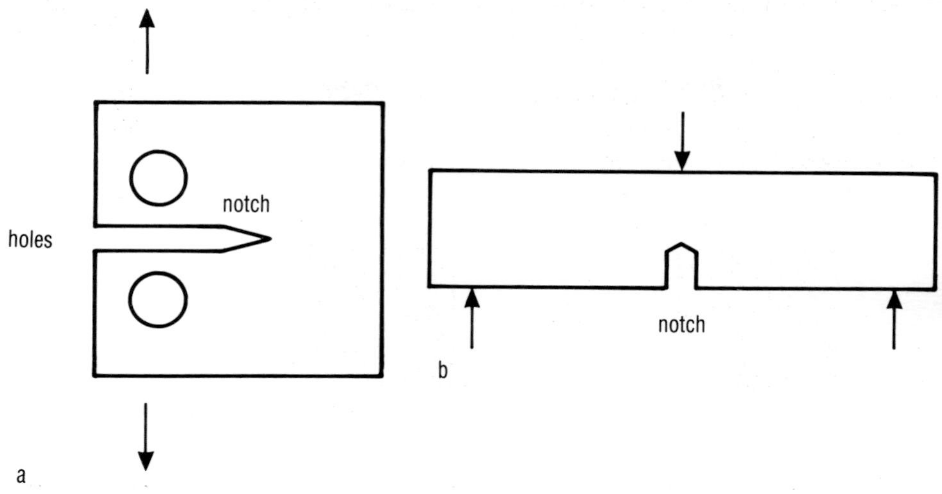

Figure 7.15

Fracture toughness of polymers and isotropic materials: (a) compact tension (holes used for loading); (b) three-point bend; (ESIS)

- for anisotropic materials *(Table 7.5)*, data normally only obtained for delamination failures between/along the ply or aligned fibre direction *(Figure 7.16)*.
- joint test protocol under development (1991) on a world-wide basis for these types of delamination failures (ESIS/ASTM/JIS).

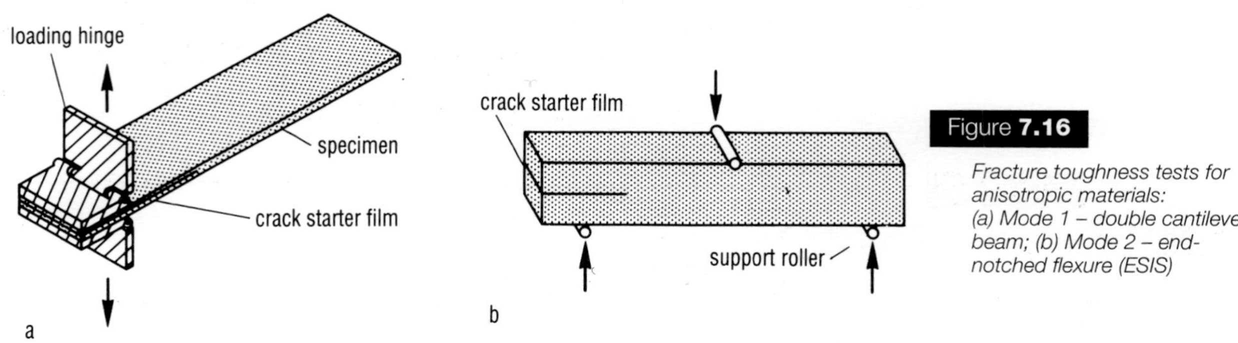

Figure 7.16

Fracture toughness tests for anisotropic materials: (a) Mode 1 – double cantilever beam; (b) Mode 2 – end-notched flexure (ESIS)

7.6 **Test methods for non-mechanical properties**

Thermal and electrical properties are important in many applications, such as the extensive use of GRP laminates in thin sheet form for circuit boards and for large components in electrical generating equipment. Several non-mechanical properties will have design implications for mechanical performance; for example, the need to protect against lightning strike in cases such as helicopter blades, and to limit the stresses developed through differential thermal expansion.

Non-mechanical property standards are listed in *Table 7.15* and fire property tests are listed in *Box 4.8*.

Table **7.15**

Non-mechanical and other properties and standards

title	number	scope
coefficient of thermal expansion	ASTM D696	(P)
thermal conductivity – heat flow meter	ISO 8301	(P)
– hot plate	ISO 8302	(P)
Shore hardness	ISO 868	(P)
Barcol hardness	EN 59	(G)
resistivity of conductive plastics	ISO 3915	(P)
electrical properties	ISO 1325	(P)
tracking index	IEC 112/587	
volume resistivity	IEC 93	
dielectric dissipation factor	IEC 250/377	
permittivity	IEC 250	
insulation resistance	IEC 167	
electric strength – power frequencies	IEC 243	

Note: IEC standards and ISO 1325 apply to 'insulating materials'.

7.7 Assessment of rate, temperature and environmental effects

Temperature and rate effects

The properties of fibre reinforced plastic materials can vary with loading rate or service temperature *(Section 3.6)*. It is therefore often necessary to repeat the standard tests over a range of suitable test conditions. Obviously some test geometries and test rigs are easier to use than others at elevated temperatures. In some tests there may be a change in failure mode (for example from tension to compression in a flexure test) and this may influence both the analysis of the test and the criteria used for failure.

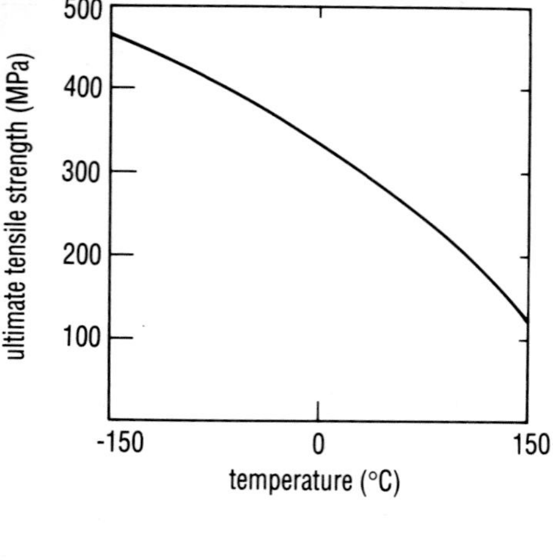

Figure **7.17**

Variation of tensile strength with temperature along a principal fibre direction for a fine weave glass fabric/epoxy laminate (Sims)

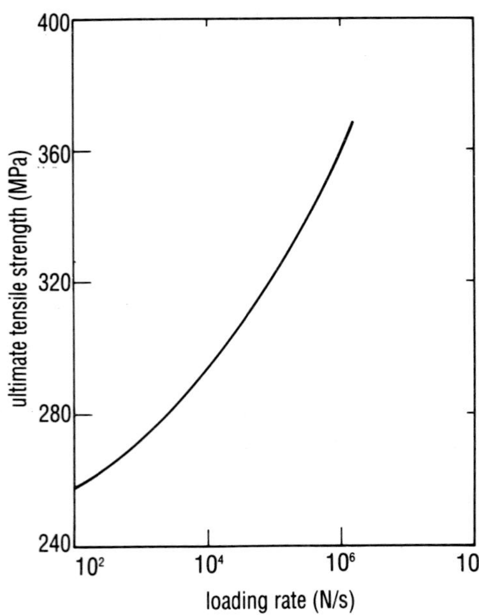

Figure **7.18**

Variation in tensile strength with loading rate for the laminate in Figure 7.17 (Sims)

Depending on the system, properties may vary in the fibre-dominated directions as well as in the matrix-dominated properties. The rate dependence observed for GRP in tension is thought to be related to the stress corrosion mechanism involved in failure of glass fibres. Tests in pure compression, or where there is a compression component such as flexure, may, even for a fully aligned system, show a dependence due to the role of the matrix in supporting the fibres against buckling. Typical trends for a GRP laminate are shown in *Figures 7.17* and *7.18*, in which the strong dependence on temperature and strain rate can be seen for a principal fibre direction.

Environmental tests

The effect of environmental factors was discussed in *Section 3.7*. Methods are available for evaluating long-term performance, but a major disadvantage in many tests is that the specimen is treated unstressed in the required environment and then tested. This underestimates the combined effect of stress and environment, particularly with flexural loads when the highly stressed surface is also the most readily attacked.

A method of testing a pipe section under the combined effects of stress, temperature and a corrosive environment is illustrated in *Figure 4.2*. If a protective liner like polypropylene is used, this test could also be used to test the effectiveness of the liner. For tensile-loaded systems the time needed to saturate the specimen with moisture or corrosive fluid, particularly if it is several millimetres thick, will reduce the degree of underestimation shown by unstressed results.

Standards relevant to environmental resistance testing are given in *Table 7.16*.

Table **7.16**

Environmental standards

title	number	scope
standard atmospheres	ISO 291	(P)
	EN* 2743	(RP)
constant relative humidity conditions	ISO 483	(P)
water absorption	ISO 62	(P)
liquid chemicals (including water)	ISO 175	(P)
demineralized water	EN* 2378	(RP)
liquid chemicals	EN* 2489	(RP)
humid atmospheres	EN* 2823	(RP)
environmental stress cracking		
– bent strip	ISO 4599	(P)
– ball/pin	ISO 4600	(P)
– tensile	ISO 6252	(P)
natural weathering	ISO 4607	(P)
damp heat, water spray, salt mist	ISO 4611	(P)
light source exposure	ISO 4892	(P)

Note: From the standard atmospheres given in ISO 291, 23°C and 50% relative humidity are normally chosen.

Methods aimed at achieving equilibrium saturation are under discussion, but there is a strong industrial demand for a fixed-time exposure to fluids. Because of the time dependence on thickness and material, however, it is difficult to select a time suitable for all materials.

7.8 **Assessment of creep and fatigue loads**

The parameters influencing the long-term effects of static loading (creep) or fluctuating loading have been discussed in *Section 3.5*. The assessment of this type of loading is not well established in standards, partly because the relative importance of these parameters is not completely understood.

The standards in Table 7.17 should therefore only be used as a guide. For example, tensile fatigue is covered by ASTM D3479, but no recommendations are given for parameters such as test frequency and loading waveform, which have been shown to influence test results for some systems.

Table **7.17**

Creep and fatigue standards

loading type	number	scope
creep		
– tensile	ISO 899	(P)
– tensile, flexure	ASTM D2990	(P)
– compressive	ASTM D2990	(P)
fatigue		
– tensile	ASTM D3479	(F)
– flexure	AFNOR T51-120	(P)

Likewise AFNOR T51-120 covers fatigue tests in flexure, and includes details of the jig designs, but not the test conditions and frequencies. An international round-robin programme on tensile and flexure fatigue tests on fully aligned systems is under way (1992) which aims to propose a preferred test method.

For reinforced plastics with a high measure of anisotropy it may be necessary to test specific properties under long-term loads in any relevant direction, such as relaxation under a compressive load in order to determine bolt torque retention. In these cases static specimen tests should be used, but with caution *(Figures 7.13, 7.14)*.

Accelerated tests

When accelerating a test by increasing the test frequency, care must be taken that autogenous (self-generated) heating does not lead to large temperature rises as a result of the poor conductivity and high damping of reinforced plastic materials. The sensitivity of some systems to a rise in specimen temperature limits the maximum frequency used (for instance, 1 – 10 Hz for GRP). Carbon fibre systems have higher frequency limits than glass fibre systems as a result of their better thermal conductivity. The test results may depend on the frequency even in the absence of this heating effect if the material is strain rate sensitive, like GRP *(Section 7.7)*.

In some cases the occurrence of microdamage (for instance, a leak test limit for containers), or a specified loss in stiffness or load rather than specimen fracture, may be used to define failure.

7.9 Finished product test methods

There are only a few standards applying to pultrusions and filament-wound systems, whereas a fairly complete series exists for sandwich laminates (ASTM aerospace series). It is likely that more test methods and performance specifications will be available during the 1990s. Complete systems such as pipes, vessels and tanks are discussed in *Section 4.5*.

Table **7.18**

Finished product test standards

product type	aspect	number	scope
rod composites	flexural strength	ISO 3597	(G)
	compressive strength	ISO 3605	(G)
sandwich constructions	flatwise shear	ASTM C273	(G)
	transverse tension	ASTM C297	(G)
	edgewise compression	ASTM C364	(G)
	flexure	ASTM C393	(G)
	flexure creep	ASTM C480	(G)
	laboratory ageing	ASTM C481	(G)
filament-wound composites	ring tension	ASTM 2290	(P)/(RP)
	cylinder, hydrostatic compression	ASTM 2586	(G)
vessels and tanks	design	BS 4994	(G)
water pipes and fittings	design	BS 5480	(G)
pipes, fittings, joints for process plant	design	BS 6464	(G)
laminated rods and tubes		BS 6128	(RP)
properties and test methods	specifications	ISO 4849	(G)
industrial laminated sheet	specifications	ISO 1624	(RP)
moulding compounds	specifications	BS 5734	(RP)

Note: sandwich structure standards apply to all materials and may contain metallic, polymeric and reinforced plastic components.

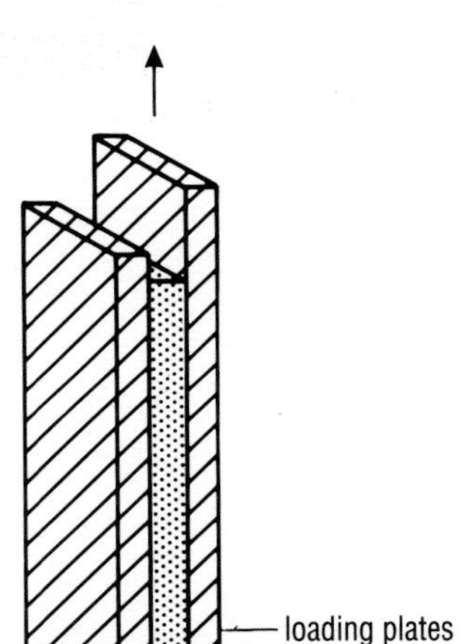

Figure **7.19**

Flatwise shear of sandwich laminate (based on ASTM C273)

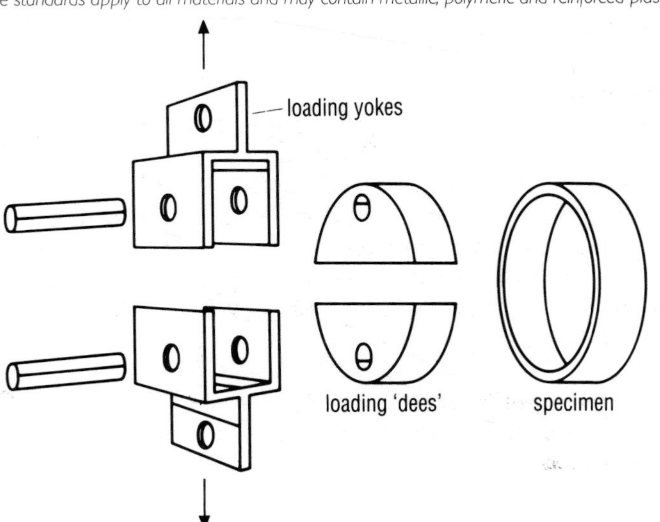

Figure **7.20**

Tension loading of filament-wound ring (ASTM D2990)

The flatwise shear testing of a flat laminate and the tensile loading of a ring specimen for two types of components in their critical loading directions are shown in *Figures 7.19* and *7.20.*

7.10 Component, structure and system testing

Component testing

It is usually necessary to test the component itself or, for large or complex systems, a sub-component such as a stiffened panel. Sub-components are chosen according to their potential criticality to the overall design and the uniqueness of any new design features or procedures.

Depending on the complexity of the component or sub-component and the service loads, this stage may require specialized test fixtures to reproduce the service loading conditions accurately. In assessing the loads to be applied, consideration must be given to the 'abuse' loads from misuse (for instance, handling of pipes during installation) as well as the normal 'service' loads.

For example, the compression after impact test used for thin aerospace panels *(Figure 7.12)* is an attempt to simulate the effect of impact loads on aircraft skins resulting from dropped tools or stones thrown up from runways.

As in the previous level of testing, a full consideration of environmental, fatigue and other conditions must also be undertaken.

Structural testing

The final stage in any development programme is to assess the final product under actual or simulated service conditions. The ease with which this can be carried out depends on the size and complexity of the component, and on the degree of knowledge of the service loads.

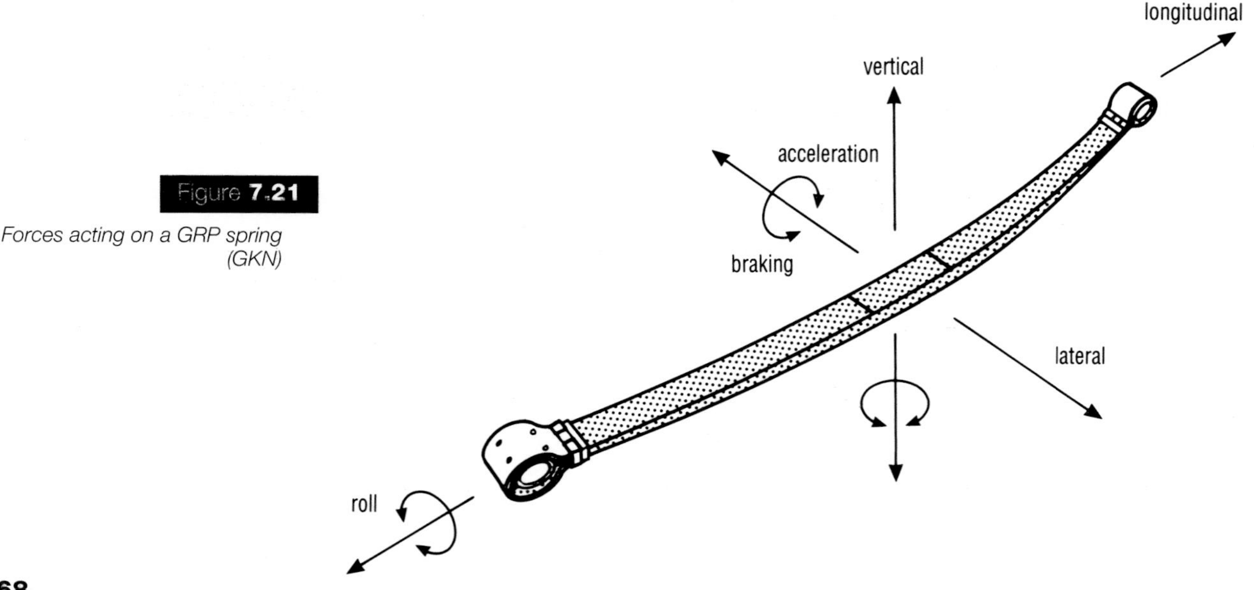

Figure 7.21

Forces acting on a GRP spring (GKN)

For example, it is possible to obtain loading profiles for a road vehicle spring by driving the vehicle over different terrains. This data can then be used to drive a multi-axial loading frame to reproduce the loading conditions accurately *(Figure 7.21)*. For larger structures such as mine-sweepers it may only be possible to test sub-elements and then to monitor the complete structure during service.

Due to the anisotropy of strength and stiffness properties of composite materials, it is important that any simulated service load faithfully represents all the loads present. For example, a transverse or shear stress significantly lower than the axial stress may be critical for a component with fully aligned fibres which will have relatively poor transverse or shear strength. Any weakness in the design due to this must be attended to by modification of the loading paths or the fibre arrangement.

Systems testing

In many cases the reinforced plastics component will form part of a larger system, for example leaf springs in vans and trucks, blades on helicopters and wind turbines, and light-weight levers in machinery. All the properties of reinforced plastics components require careful consideration as interfacing them with other systems is not always straightforward.

For example, if a component is bolted to another member, its design must avoid damage due to the applied 'crushing' load; in the case of a leak-tight component, there must be no unacceptable loss of bolt torque which results in fluid leaks. Although these aspects are studied at the coupon or component level, additional inspection will be needed during system testing. Testing should use the expected or actual service conditions of the completed system.

7.11 Quality assurance and non-destructive evaluation

One essential requirement for quality assurance and quality control *(Section 4.5)* is the non-destructive evaluation of a component or structure. The available inspection techniques can be grouped into those which require minimal equipment and those which imply significant capital investment, as shown in *Tables 7.19* and *7.20*.

Table **7.19**

Low-cost inspection techniques

method	standard	comments
fibre weight fraction	–	proportion of reinforcement to resin used
cure cycle	–	temperature/time profile of resin cure
Barcol hardness	EN 59	micro-hardness test, degree of cure of resin
component mass	–	process control
visual inspection (quality control)	BS 4549	particularly good for transparent and translucent materials, indicative of moulding quality
dimensions	–	process control
resonance	–	mechanical tap test, detects certain types of defects
die penetrant	–	detection of surface cracks

Note: refer to Table 7.14 for other methods for fibre content

Low-cost inspection technology

The level of testing will depend upon both the degree of automation in the manufacturing process and the structural loading to which the component will be subjected. The techniques set out in *Table 7.19* can be readily undertaken within the manufacturing area of a plant.

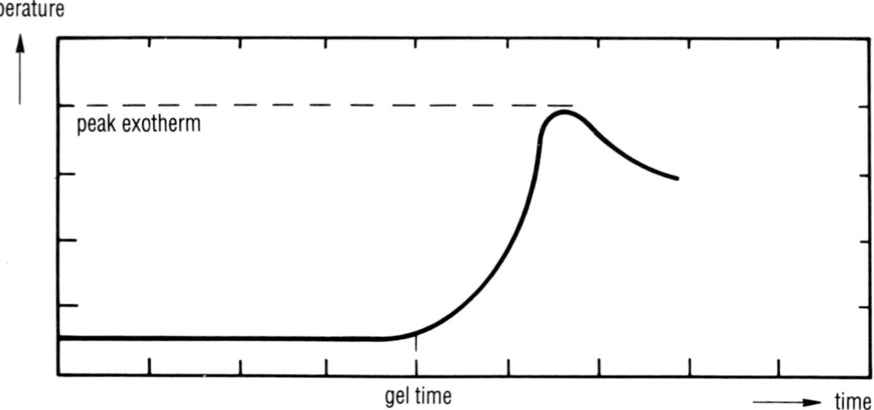

Figure 7.22

Exothermic temperature rise when a thermosetting resin begins to harden. The gel time and peak exotherm are possible process control parameters (Scott Bader)

Fibre weight fraction can be determined by measuring the quantity of reinforcement and resin used for each moulding. This is particularly important for large mouldings. For more accurate tests refer to *Table 7.14*.

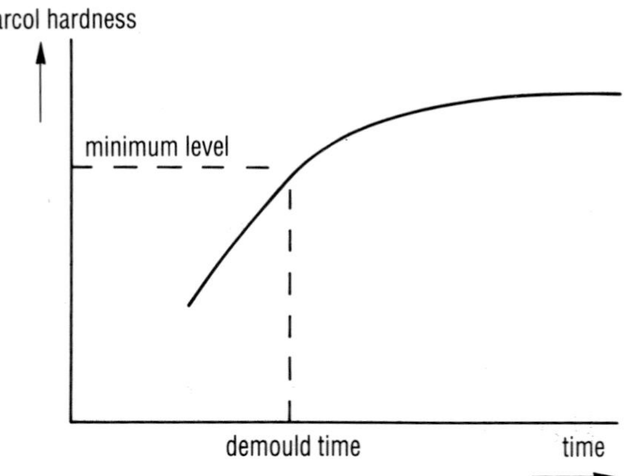

Figure 7.23

Typical increase in Barcol hardness with time after a thermosetting resin has been catalysed. Setting a minimum hardness level will provide the time at which a component can be demoulded

The stages of **cure** for a thermosetting resin are set out in *Table 5.7*, and each can be used to provide process control. For example, when a catalysed resin starts to harden, heat is evolved over a period of time, as shown in *Figure 7.22*.

Barcol hardness is widely used to determine the degree of cure of a resin, as it increases with time and eventually reaches a maximum value *(Figure 7.23)*. Its value after a given time can be used to decide when to demould a component. Since it uses a hand-held micro-hardness indenter, the method can be readily used to check components whatever their size.

Component mass can be determined by weighing and setting prescribed limits.

Visual inspection is a good check of the quality of the moulding, which should be free from protruding fibres, voids, pits, bubbles, cracks, blisters, resin-rich or resin-starved areas and foreign matter. If the component is translucent or transparent then its entire volume can be inspected. As well as showing up processing faults, surface appearance is very important if a good finish is required, for example in boat hulls and automotive panels.

Dimensional checks are best made using templates or gauges, and are a good indication of the level of process control. Thickness, which gives a good indication of the degree of consolidation if the reinforcement is in layers, can be measured with a portable ultrasonic gauge in most industrial settings.

Resonance induced by tapping with a metal object such as a coin is indicative of the quality of the moulding, particularly if internal defects such as delamination are present *(Figure 7.7)*. A hand-held instrumented gauge is now available.

An example of the techniques adopted for assessing the quality of a wind turbine blade is given in *Box 4.4.*

Table **7.20**

Specialized inspection techniques

method	comments
acoustic emmission	detects damage formation as load is applied or during service (ASTM E1067, ASTM E1118)
thermography	remote technique capable of detecting surface damage or underlying damage during service
ultrasonic	locates defects like cracks and delaminations
radiography	same methods used as for metals

Specialized inspection techniques

The more specialized techniques are listed in *Table 7.20*. The techniques vary in their ability to detect damage, and expert advice should be sought as to the relevance of a technique for a specific application.

Acoustic emission detects the noise emitted when damage occurs on loading by using transducers (as receivers) coupled to the surface of a component or structure. Damage can be located if a transducer array and suitable timing equipment are used. A procedure has been developed for proof-testing pressure vessels and storage tanks.

Thermography uses an infra-red detector to monitor variations in thermal emission which occur at cracks or damaged areas at or near the surface. The thermal emission may be either applied or self-generated, because of the poor thermal conductivity of reinforced plastics. The detector can be remote from the component.

Ultrasonic scanning involves scanning the component with an emitter of ultrasonic waves and a detector, or an emitter/receiver, to detect the presence of a defect *(Figures 7.24 and 7.25)*. Good coupling is required between the component and the emitter/detector.

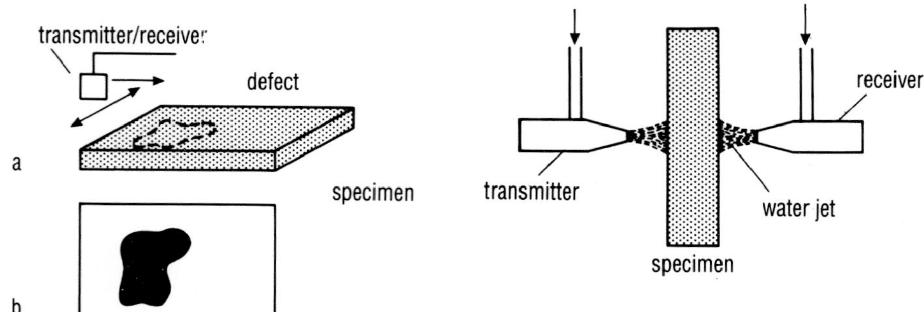

Figures **7.24 & 7.25**

Ultrasonic detection of a defect: (a) the transmitter/receiver scans the specimen in a water tank; (b) c-scan image of defect (Clarke)

(c) Use of water jets for coupling ultrasonic probes to a large component (Clarke)

Several recent reviews cover these techniques in more detail, as well as others such as chemical spectroscopy, corona discharge, eddy current and dielectric methods. In specific cases, computer processing of images is being used to enhance the quality of the information gained. The area scanning methods such as acoustic emission are best used in conjunction with local area techniques.

7.12 Reference information

General references

Standard bodies

International Standards Organization (ISO) – standards available from national bodies.

Comité de Européen Normalisation (EN) – standards available from national bodies.

British Standards Institution (BSI), 2 Park Street, London W1A 2BS, UK.

Association Français de Normalisation (AFNOR), Tour Europe, La Defence, Cedex 7, 92080 Paris, France.

Deutches Institut für Normung eV (DIN), Postfach 1107, D-1000, Berlin 30, Germany.

American Society for Testing and Materials (ASTM), 1916 Race Street, Philadelphia, PA 19103, USA.

Japanese Industrial Standards (JIS) c/o Standards Department, Agency of Industrial Science and Technology, Ministry of International Trade and Industry, 1-3-1 Kasumigaseki, Chiyoda-ku, Tokyo, Japan.

Complete copies of British and International standards can be purchased from BSI Sales, Linford Wood, Milton Keynes MK14 6LE, fax +44 0908 320 856.

Other Sources

Special Technical Publications (STPs) published by ASTM (see above) – good source of reviewed conference papers on topics such as 'Composite Materials –Testing and Design' and 'Fatigue and Fracture'.

Handbook of Plastics Test Methods, R Brown (ed), George Godwin/PRI, London, 1981 – particularly strong on plastics test methods.

Experimental Characterization of Advanced Composite Materials, L A Carlson and R B Pipes, Prentice-Hall, New York, 1987 – good introduction to principles and methods for composite test methods.

Manual on Experimental Methods for Mechanical Testing of Composites, R Pendleton and M Tuttle (eds), Elsevier, Barking, 1989 – covers experimental techniques for testing composites.

VAMAS – a collaborative agreement involving a large number of countries in Europe, North America and Japan in developing material test standards. Contact Graham Sims, National Physical Laboratory, Teddington, Middlesex TW11 0LW, UK.

Specific references

7.1 Introduction

Development of Standards for Advanced Polymer Matrix Composites – a BPF/ACG overview and proposal, G D Sims, Report DMM (A) 8, NPL, Teddington, 1990.

7.2 **Material anisotropy**

'Effect of test conditions on the fatigue strength of a glass-fibre laminate: Part A – Frequency', G D Sims and D G Gladman in *Plastics and Rubber Materials and Applications* 1, 41–48, 1978.

Composite Design, S W Tsai (3rd edition), Think Composites, Dayton, 1987.

7.3 **Pyramid of substantiation**

'Secondary source qualification of carbon fibre/epoxy prepregs for primary and secondary Airbus structures', K Schneider and R W Long, *Plastics, Metals, Ceramics Proceedings 11th SAMPE Conference*, Basel, 1990.

7.4 **Fibre, matrix and intermediate materials test methods**

'Specifications, test methods, and quality control of advanced composites', J M Methven in *Design With Advanced Composite Materials*, ed L N Phillips, Design Council, London, 1989.

7.5 **Coupon test methods**

'Standardization of glass fibre products', in *Fibreworld*, Vetrotex, Chambery, June 1989.

European Structural Integrity Society (ESIS), previously called the European Group on Fracture (EGF), is involved with developing standards on fracture; secretary is Paul Davies, IFREMER, Centre de Brest, BP 70, 29280 Plouzané, France.

CRAG Test Methods for the Measurement of the Engineering Properties of Fibre Reinforced Plastics, P T Curtis (3rd edition), Technical Report TR 85099, RAE Farnborough, 1989.

Computer Procedures and Experimental Study of Fibre Reinforced Composites, G D Scowen, Report No 668 NEL, East Kilbride, 1980.

'Strain gauges on glass fibre reinforced polyester laminates', S J Thompson, R T Hartshorn and J Summerscales, *Proc 3rd Conf on Composite Structures, Paisley*, Elsevier, London, 1985.

7.8 **Assessment of creep and fatigue performance**

A VAMAS Round-robin on Fatigue Test Methods for Fibre Reinforced Plastics: Part 1 – Flexural and tensile tests of unidirectional material, G D Sims, Report DMA (A) 180 NPL, Teddington, 1990.

7.11 **Quality assurance and non-destructive testing**

Non-destructive Testing of Fibre-reinforced Plastics Composites, Vols 1 and 2, J Summerscale, Elsevier Applied Science, London, 1987, 1988.

'Non-destructive evaluation of composite materials', B Clarke in *Metals and Materials* 6, 135 – 138, 1990.

Non-destructive Evaluation of Composite Structures, D E W Stone and B Clarke – an overview of 6th ICCM 2nd ICCM, London, Elsevier, Amsterdam, 1987.

'The quality control and non-destructive evaluation of composite materials and components', P R Teagle in *Design With Advanced Composite Materials,* ed L N Phillips, Design Council, London, 1989.

Footnote

This chapter was prepared by Graham Sims, National Physical Laboratory, Teddington, UK.

Design studies

chapter 8

summary

This chapter provides four brief design studies which highlight different stages and aspects of designing with reinforced plastics, following the methods outlined in the earlier part of the book.

The golf club study is conceptual only; the road wheel illustrates the constraints of replacement design; the suspension arm demonstrates how new concepts can be developed; and the tennis racket study shows how imaginative processing and design can allow comparatively expensive materials to be used in marketable products.

Introduction

Two of the studies are from work undertaken at the National Engineering Laboratory for a consortium of British companies to design replacement structural automotive components using reinforced plastics such as a road wheel, trailing suspension arm, coil spring and clutch housing.

The aim was to demonstrate the feasibility of, and build up confidence in, using this group of materials for such highly stressed components. No price premium was likely to be paid for reductions in mass as the cost of mass-production vehicles is very price sensitive.

Sports goods are different, however, as performance is more important to many players than cost. So potential improvements in performance were the driving force for the other two studies.

8.1 Golf club shaft

This study shows how the principal loadings of a golf club shaft are resolved so that an initial design can be undertaken.

Design brief

The requirement was to replace the steel shaft of a golf club with one manufactured from fibre reinforced plastics, and to identify a suitable and cost-effective manufacturing process for volume production.

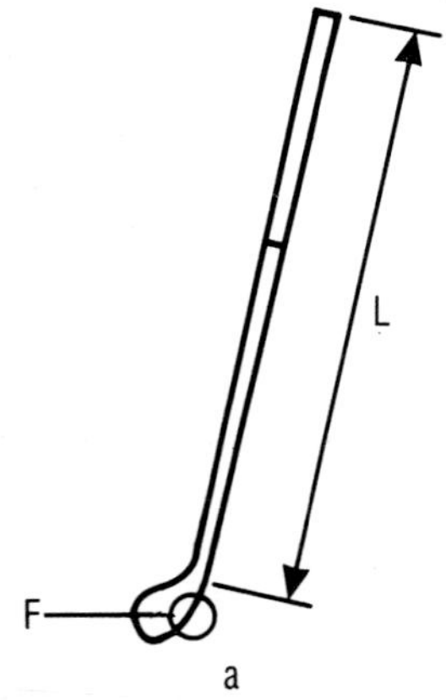

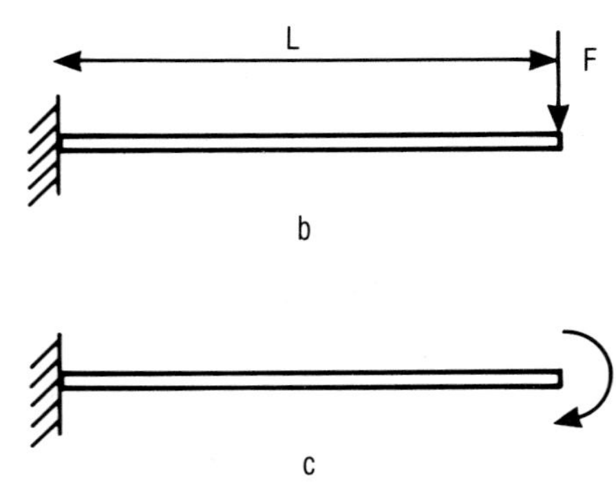

Figure **8.1**

Simplified modelling of forces and loadings on a golf club shaft at the moment of impact on the ball: (a) loading on shaft; (b) flexural loading; (c) torsional loading. The resultant moment is equal to the flexural moment plus the torsional moment

Design considerations

It was assumed that the stiffness of a steel shaft would provide an initial performance specification for the replacement shaft. The maximum loading on a golf club shaft occurs when the club head hits the ball, and is therefore a dynamic transient load. This introduces two main types of loading, namely flexural and torsional as shown in *Figure 8.1*.

A full and detailed analysis of the loads and stresses is well outside the scope of this case study, and for simplicity was assumed to be non-critical. The golf shaft was therefore assumed to be a stiffness-critical application, and the design criterion was stated as:

$$\text{steel shaft stiffness} = \text{composite shaft stiffness}$$

Since both flexural and torsional loadings are involved, each of these stiffness characteristics would need to matched.

Material considerations

Reinforcement fibres need to be inserted in the direction of the principal load directions for maximum strength and stiffness. For flexural loading this means that the fibres must lie along the axis of the shaft *(Figure 3.8)*, while for torsional loading, the optimum fibre angle is ±45° to the axis *(Figure 3.5c)*.

Neither of these two preferred fibre alignments suited both loadings, nor did they provide an acceptable stiffness for the other loading mode. So, in order to satisfy both stiffness criteria, it was assumed that some combination of axially aligned and 45° helically wound fibres was required.

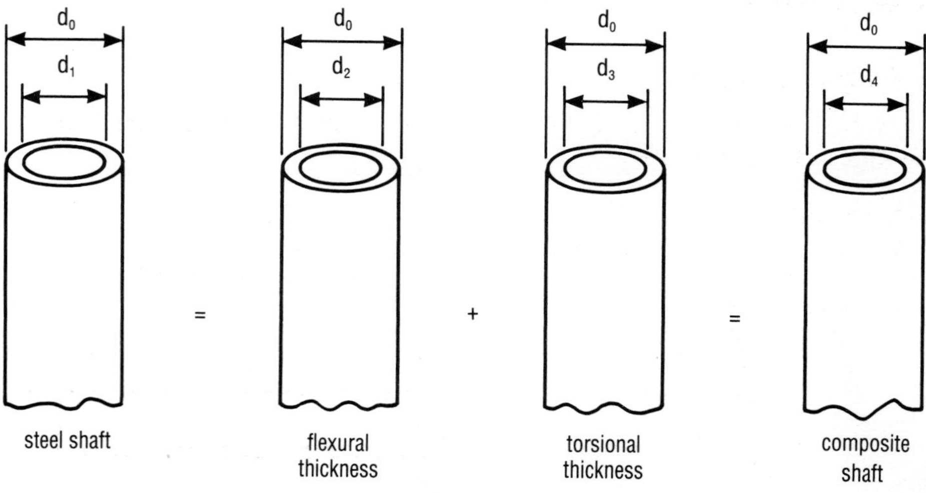

Figure **8.2**

Schematic representation of a composite shaft to match the stiffness of a steel shaft

Design calculation

The first estimate for the shaft design was therefore obtained from the equivalence of the flexural stiffness – that is, the product of the flexural modulus E and the moment of inertia:

thus $$E_S \times (d_0^4 - d_1^4)_S = E_C \times (d_0^4 - d_2^4)_C$$

where s indicates steel, c composite and d the diameters as shown in *Figure 8.2*.

A similar equation has to be satisfied for the torsional stiffness, G being the shear modulus:

$$G_S \times (d_0^4 - d_1^4)_S = G_C \times (d_0^4 - d_3^4)_C$$

Using published data, the flexural and torsional wall thicknesses can then be summed to provide a first estimate of the composite shaft thickness and hence the mass. Since steel golf clubs have either tapered or stepped diameter shafts, a series of estimates was then compiled to give an initial indication of the stiffness variations along the shaft length.

While the ratio of longitudinal to torsional stiffness of steel is fixed, it can be varied in composites by altering the ratio of axial to hoop fibres, so a straight copy may not provide the optimum design.

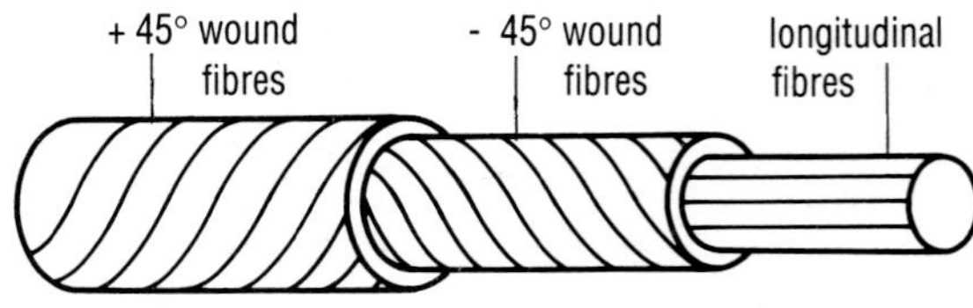

Figure **8.3**

Configuration of an envisaged composite shaft

Figure **8.4**

Golf club with a composite shaft (John Letters)

Manufacturing considerations

The manufacturing process had to be capable of laying down the reinforcing fibres, evenly dispersed around the shaft circumference, for both longitudinal and helical alignments. Prepreg materials *(Box 5.8)* would provide the necessary fibre alignment and orientation and enable the fibre/resin ratio to be controlled. Wrapping the prepreg around a mandrel would give the required shape *(Section 6.9)*.

To optimize the fibre lay-up, however, the following aspects needed some further thought:

- for a given geometry, stiffness can be varied by suitable selection of the fibre type *(Figure 5.1)* and there will be a trade off between stiffness and cost of the fibre *(Figure 2.15)*.
- if the fibres are laid down at a constant rate along a tapering shaft, some increase in wall thickness will result as the shaft diameter decreases.
- if the helically aligned fibres are laid down at a constant pitch on a tapering shaft it will not be possible to maintain their 45° alignment.
- if the different fibres are laid down in discrete layers to give a laminated shaft then, while it may be possible to maintain a uniform wall thickness, the possible ratios of longitudinal to helical fibres will relate directly to the possible combinations of discrete layer thicknesses.

Layer lay-up

It would be sensible to use the helical layers to contain and consolidate the layers of axially aligned fibres. The helical fibre pattern on the outer surface could even be considered as a selling point *(Figure 8.3)*.

Further design calculations

Whether the tube is stepped or tapered, the resultant stiffnesses of the shaft will need to be recalculated taking account of the exact positioning of the differently aligned layers within the wall thickness. Some further tuning might then be necessary before the first prototype shafts are manufactured, and the design calculations would need to be confirmed by experimental testing.

Conclusion

Golf club shafts could be manufactured from reinforced plastics and there are various options available to the designer to tune the performance of the club.

Commercialization

Current world production of carbon fibre reinforced golf club shafts is now estimated at nearly 10 million a year. This shows that sports goods purchasers will pay more for a product with a performance advantage.

The demand has been driven by the ability to tailor the properties of the shaft materials to improve performance. For example, because graphite shafts can be lighter than steel (by as much as 50% with certain designs), an equivalent amount of effort can produce a higher club head velocity.

In addition, the very rapid rate at which a graphite shaft recovers from flexing can keep the club head in contact with the ball longer. This combination will give greater distance to the drive provided that the ball is correctly struck.

The ability to vary the torque (that is, the tendency of the shaft to twist under impact), the swing weight and flex point of the shaft have all helped to boost the sales of such shafts.

8.2 **Road wheel**

This study outlines the design development of a road wheel, one of a number of individual components designed, manufactured and track tested by the National Engineering Laboratory.

Design brief

The prime requirement was for a replacement design of an existing road wheel so that that it could be fitted to a car and road tested. As with the suspension system described in *Section 8.3*, the manufacturing technique had to be compatible with industrial production rates. Also the design had to be cost-effective, and lighter than conventional steel wheels.

Design constraints

The requirement for a direct replacement restricted the available space envelope and meant that the composite wheel had to have the same large offset as an existing wheel to give the necessary track. The large offset was a significant factor in the design and contributed to the high in-service loadings that the wheel would have to withstand.

A further constraint was that the wheel should conform to BS AU50, which specifies the profile of the wheel *(Figure 8.5)*. The key performance requirements were obtained from this specification and the German Traffic Licensing Regulations. The wheel had to conform to the regulations and pass the rig and performance tests set out in *Table 8.1*.

Table **8.1**

Test requirements for a road wheel

test requirements

- inflation
- rotating bending
- simulated kerb-impact
- nut torque-retention
- simulated braking tests
- simulated cornering
- alpine brake-fade
- rolling road

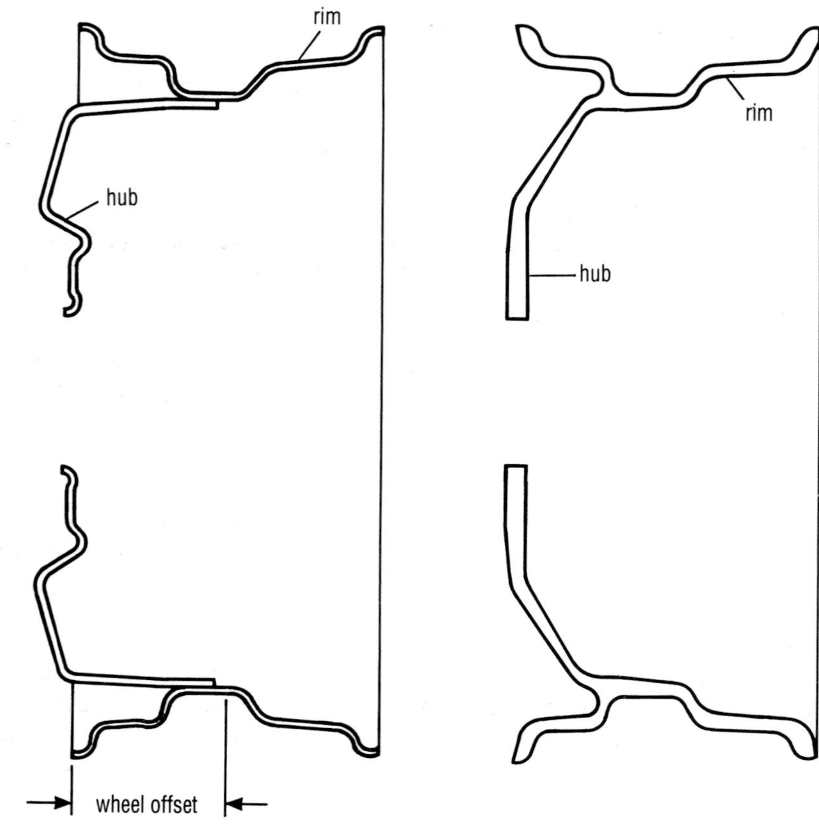

Figure **8.5**

Profile requirement of metallic road wheel made from two separate components, the hub and the rim (BS AU50)

Figure **8.6**

Profile of composite road wheel in which the hub and rim are moulded in one piece (NEL)

After passing these tests, the wheels then had to pass the stringent on-car trials under test track conditions, as well as more mundane tests like multiple tyre changes.

Design considerations

A number of designs were initially examined, taking account of the functional requirements, space restrictions, service loadings and – perhaps the most significant factor – the production volume.

As production volumes would be high, of the order of 500,000 a year, production rates would significantly influence final component costs. Possible production routes therefore had an early influence on design choices and decisions. Component geometry and symmetry were both considered best suited to a moulding route *(Section 6.3)*. Because of the high loadings and high temperatures to be met, materials with poor temperature and strength properties were immediately excluded *(Section 5.3)*.

Structural design calculations

Finite element models of several basic wheel designs were examined *(Section 3.3)* and a final wheel shape was selected from the indicated stress levels and distributions. These initial models assumed a constant component thickness.

At this stage with a conventional steel wheel, the options available to 'tune' the design to meet strength or stiffness criteria are to use either a higher strength steel or a thicker section.

However, with fibre reinforced plastics two further options are possible:

- within the same space envelope, the material properties can be improved by using higher levels of aligned reinforcement *(Figure 6.8)*.
- thickness variations can be much more readily incorporated into a moulded component than they can with the nominally constant sheet thickness of a steel wheel.

Both options were used in the design of the composite wheel.

The rotating bending test *(Table 8.1)* was considered the most severe loading condition experienced by the wheel, so it was used to set the stress limits in the final design analysis.

The maximum design stress was based on the available material properties modified by considerations of service conditions. The rotating bending test imposed a requirement of 150,000 loading cycles. Available information on normalized fatigue data indicated that a factor of 0.46 should be applied to the tensile strength of the selected material to give the allowable fatigue stress *(Section 3.5)*.

The design loop *(Figure 2.4)* consisted of progressive iterations of the wheel cross-section, modifying the thickness locally to reduce stress levels below the allowable value. The final cross-section used for the wheel is shown in *Figure 8.6*.

Material considerations

Within the constraints of the space envelope, it was evident that only a hybrid construction comprising a mixture of continuous and random glass fibres would meet the functional and cost targets. The ideal material had to combine the quick cure properties of a sheet moulding compound *(Section 5.4)* with the strength of an aligned material *(Section 2.8)*.

A continuous aligned material was produced (*Figure 6.6*) which was subsequently plied at 90° to form the stock material for the wheel's outer layers while conventional SMC was used for the core. This material was chosen partly for its low cost, reflecting the importance which the automotive industry places on price.

Manufacture

The choice of processing method centred around the production volume requirement. On the basis of cycle time, tooling costs, tool life and other factors, the only practical manufacturing route appeared to be hot press moulding *(Section 6.3)*.

While the material development programme was under way, tooling was manufactured which on completion was installed in a 250 tonne moulding press *(Figure 8.7)*. Following a series of trials to commission the tooling, work began using the design material.

The trials quickly showed up a problem not encountered on moulding the trial test pieces. The high speed of cure, the thickness of the moulding and the high exotherm combined to

Figure **8.7**

Tooling in place on moulding press (NEL)

Figure **8.8**

Wheel under test on vehicle (NEL)

produce cracking in the centre of the wheel moulding during cure.

Several attempts to solve this problem were made before it was finally solved by changing the resin system from a polyester to a vinyl ester *(Section 5.3)*. This gave the added advantage that the final wheel was slightly tougher than it would otherwise have been.

The moulding charge was prepared by building up the material of the inner and outer wheel shapes on wooden formers. The aligned material was orientated so that the fibres lay generally in the direction of the principal hoop and radial stresses, as predicted by calculation. The two halves of the charge were then removed from the formers and assembled with the core material, the whole charge being held together with a belt of material to form the rim.

A cure time of ten minutes at 150°C was allowed for each moulding. Deflashing was carried out immediately after the components were removed from the mould, and structural integrity assessed by various non-destructive evaluation tests *(Section 7.11)*.

Testing

A comprehensive series of rig tests and road trials was carried out following *Table 8.1*. The wheel is shown under test on a vehicle in *Figure 8.8*. In addition, full road trials were successfully conducted at a variety of sites and under various conditions.

The only test which showed up inherent problems was the alpine brake fade test which, due to its severity (the brake pads glow red hot), suggested a need for further material development or some additional heat shielding on the appropriate wheel surface.

Conclusion

The design brief was fulfilled in that it was shown to be possible to design and manufacture a road wheel from fibre reinforced plastics by a mass production technique. A saving in mass of up to 50% of the pressed steel version was achieved. The development programme showed that a composite wheel could meet all the regulations and pass all the required tests.

Commercialization

Parallel work done by Goodyear Tyre and Rubber Company has reached a stage at which a small number of high performance vehicles were fitted with composite wheels in 1989.

The material and manufacturing route used are similar to that described above, though the shape of the spokes is very different *(Figure 8.9)*. This composite wheel offers a 10 – 15% reduction in weight over a cast aluminium wheel and is some 30 – 45% lighter than a mild steel wheel.

Other claimed benefits are that it resists corrosion, that it is more uniform from wheel to wheel and has improved run-out and balance compared with a steel wheel. Moreover it can be moulded in one piece. No additional machining is needed to make the wheel round and only the bolt and valve holes have to be drilled.

8.3 **Novel car suspension system**

The replacement design of individual components can never be fully effective because of the restrictions imposed by the dimensional envelope and the fixing points to which such components must be attached. The rear suspension system of a vehicle was therefore designed using a 'clean sheet of paper' approach so as to use the inherent advantages and potential of fibre reinforced plastics to the full.

Figure **8.9**

All-plastic wheel manufactured by Motor Wheel

Design brief

The brief was to design a new fibre reinforced plastics rear suspension system for a family car to determine and exploit any particular advantages of composite materials in the suspension system, and to investigate the following design aspects:

- opportunities for parts consolidation to reduce assembly time.
- mass reduction, particularly of unsprung mass.
- manufacture of a prototype system by a route which would be feasible for volume production.
- fitting the prototype to a production vehicle and track testing.

Conceptual design

The principal components of the new suspension system consist of two flexible twin-leaf springs or arms coupled by a cross-member which forms the rear wheel axle *(Figure 8.10)*. Each arm is mounted in a three-point bend configuration attached at one end to the car body, and at the other to the wheel axle. There is also an intermediate support point reacting against the underbody of the car.

The cross-member has enough structural rigidity to maintain wheel alignment under maximum loading conditions, but allows the wheels independent vertical movement because of its low torsional stiffness.

Design considerations

Throughout the design, a number of aspects were considered in the light of the specific needs of the manufacturer and the end-user *(Table 8.2)*. These aspects varied in priority at different stages in the design cycle *(Figure 2.3)*.

The suspension system of any car must also cope with a number of different loading configurations in service *(Figure 8.11)*. In addition, certain dimensions had to be maintained since the system had to be fitted to a vehicle in order to be road tested. The existing track width, wheel size etc were therefore kept the same as that of the car to which it was to be fitted.

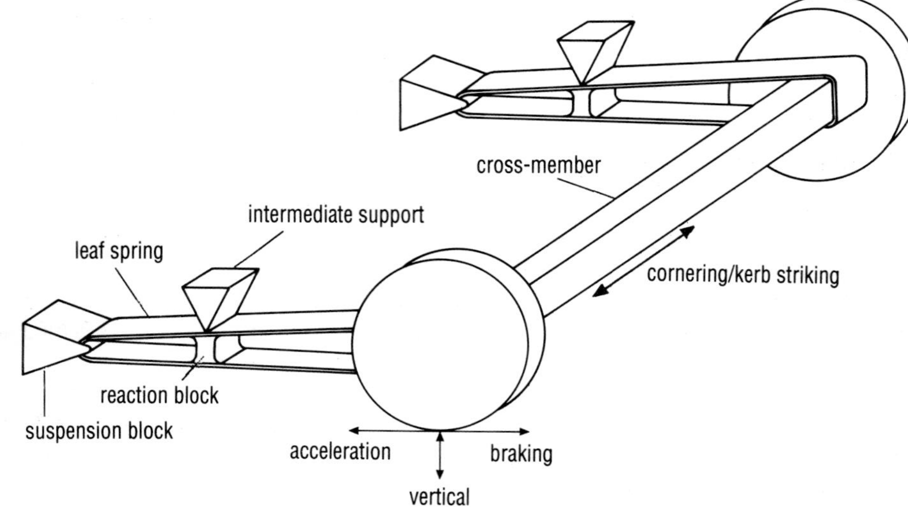

Figure 8.10

Principal components of the redesigned suspension system and the loadings acting upon it

Table 8.2

Design considerations of novel car suspension

design guidelines
- performance specification of existing system
- dimensional envelope, existing and potential
- materials (fibres/resins)
- manufacturing route
- cost
- parts consolidation
- weight reduction
- design for production
- detail design

An MG Maestro 2.0 EFI was chosen for the following reasons:

- it was a typical example of a standard sized family car.
- it was relatively easy to modify.
- its high-performance engine meant that the car could be tested in extreme cornering conditions.
- a detailed specification existed.

Component considerations

The performance specification for the fibre reinforced plastic suspension was taken to be that of the all-steel design.

Spring The relatively low modulus of elasticity of glass fibre reinforced epoxy compared with that of steel *(Table 2.7)* enabled the twin leaf arms to perform the function of the conventional coil spring in terms of vertical rate and displacement within a reasonable and acceptable length.

Also, the requirements for the car's roll and bounce compliances could be accommodated by loading the springs in a three-point bend configuration *(Figure 8.11)*. A steel leaf spring of similar design would be too long to package within the car body because of its much higher elastic modulus. This design also allowed the assembly to be relatively rigid against forces other than those normal to the road surface, so the wheel deflection design tolerances, such as toe and camber angle, were met. In fact, the replacement design performs two separate functions:

- as a spring to accommodate the vertical compliance and travel.
- as a structural member replacing the conventional trailing arm.

Careful consideration of the maximum fibre stress within the spring had to be made because a component of this type is fatigue critical. A realistic value of 600 MPa was assumed, obtained from published literature related to GRP leaf springs.

Limiting the maximum fibre stress to this value allowed the thickness of the leaf spring to be determined. Due to the large numbers of parameters involved and the complexity of the displacement and the stress fields induced by the loading, an iterative rather than a closed form analysis procedure was used to design the size of the structure.

The **cross-member** rigidly connects the free or axle end of the two twin-leaf springs *(Figure 8.10)* and performs the following functions:

- it provides a stiff rigid beam in bending so as to maintain the wheel camber and toe deflections within the design specification.
- it contributes a specific value to the overall roll stiffness of the car by having a relatively low torsional stiffness.
- it provides an easy and effective means of attaching the two springs together and to the wheel bearings.

This balance of properties is not so readily obtained using metals. Reinforcement needs to be provided along the principal stress directions *(Section 3.2)*. In this way the beam can be designed to provide the desired combination of flexural and torsional stiffness.

A **reaction block** was fitted between the two leaves of the spring at the reaction point where the spring is supported by the car body via an intermediate support *(Figure 8.11)*. The block forms a spacer to keep the two spring leaves apart and has to cater for the

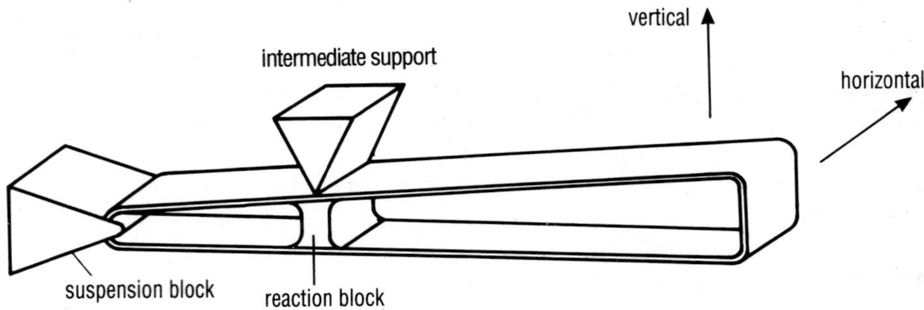

Figure 8.11

Leaf spring showing loading at three points: the suspension block, the reaction block and at the end of the spring. The spring is compliant vertically but rigid horizontally (NEL)

differential movement between the two leaves that occurs as the spring bends. It therefore has to be very stiff in compression and flexible in shear. This was achieved by bonding together thin layers of reinforced plastic and elastomeric material.

Materials considerations

Spring Principal factors were the cost, and the level of the elastic modulus necessary to meet the design specification, as fibres such as carbon were too expensive *(Figure 2.15)*. The modulus could be achieved by using a high glass fibre content, unidirectional material. As the component was also fatigue limited, it was considered necessary to use an epoxy resin system as the matrix *(Section 5.3)*.

Cross-member The complex loading of the cross-member necessitated various types of reinforcement. Fibres were laid along the length of the beam to provide maximum bending stiffness (*Figure 3.8*), and similar continuous fibres in a cross-plied or angular alignment to the longitudinal axis of the beam to provide a 'tuned' torsional stiffness *(Figure 3.5)*. Both unidirectional glass fabric as well as chopped strand mat *(Section 5.2)* were incorporated in the fabrication process to achieve the desired combination of flexural and torsional stiffnesses.

Manufacturing considerations

Leaf spring The required fibre alignment along with the final shape, an elongated closed loop, made this an ideal component to be produced by filament winding *(Section 6.6)*. It could then be sized to the final dimensions using matched tooling *(Section 6.3)*. This process consolidated the fibres over the flat surfaces and achieved the necessary thickness control to limit fatigue stresses.

The **cross-member**, being a channel section, was an ideal shape to be made by pultrusion *(Section 6.7)* because of its uniform section and long length. However, because of the expense of a dedicated pultrusion die, prototypes were fabricated by contact moulding using hand lay-up techniques *(Section 6.2)*. The prototype suspension system is shown in *Figure 8.12*.

Testing

The computer predictions of spring rate and contribution to roll stiffness of the springs and the cross-member were independently checked on test rigs prior to assembling the complete system. Fatigue testing was also carried out on material cut from prototype springs.

The vehicle chassis was extensively modified, mainly to provide new anchorage points and some additional space to enable the suspension to be fitted and road tested. Road testing at the Motor Industries Research Association test track was carried out by experts from one of the consortium companies. Three variations of the system were built and tested to enable more information to be gathered regarding the use of FRP materials for this type of structure. This information and data were collected from static deflection, subjective handling and noise measurements.

Conclusion

A complete rear suspension system for a car was successfully designed and built. Although only prototypes were tested, the results were very promising. In particular the drivers'

Figure **8.12**

Rear wheel suspension system suitable for fitting to an MG Maestro family car (NEL)

subjective views were that the twin-leaf system was comparable to the original system fitted to the car. The design objectives were met in that it had been possible to consolidate the number of parts and manufacture a prototype by a route amenable to volume production.

8.4 **Tennis racket**

This study outlines the successful application of innovative fabrication technology to develop a new type of tennis racket for the sports goods market.

Design brief

The prime requirement was to develop a novel tennis racket which had improved playing characteristics, performance and weight characteristics over any existing type of racket. A further requirement was the process should be automated, with a low labour content.

Design considerations

Basic structural and performance considerations dictated a frame capable of withstanding:

- a tensile loading of 300 N in each of 20 strings.
- a superimposed impact loading associated with hitting the ball.
- a frame weight in the range of 220 – 260 g.

Further analysis indicated some additional requirements:

- a hollow cross-section to reduce mass
- internal hollow pillars at stringing points to provide additional strength.
- a foamed polymer core of specified density to control mass and balance and provide

good vibration damping.
- a good quality surface finish.

Materials considerations

Existing rackets consisted either of laminated wood or prepreg mouldings of glass or carbon fibre fabrics *(Section 6.8)*. The high price differential between wood and carbon fibre *(Figure 2.14)* meant that the more expensive materials had to be used as efficiently as possible.

Structural calculations indicated that the same stiffness could be obtained with a material of 20% lower strength and a larger, hollow cross-section *(Box 2.6)*. Attention therefore focussed on the use of short carbon fibres in a polyamide (nylon 66) matrix. This material was considerably cheaper (about £15/kg) than carbon fibre prepreg, though still 30 times more expensive than wood.

Manufacturing process

The target annual volume production was about 40,000 to 120,000 rackets. These large quantities *(Figure 2.11)* suggested the use of injection moulding *(Section 6.9)*, which would enable the racket to be moulded direct to final dimensions with a good quality finish. The chosen material was suitable for this production process.

However, it took considerable ingenuity and engineering development to enable the nylon/carbon fibre composite to be formed precisely to the complex geometric specification while retaining the necessary material properties. The hollow core was achieved by injection moulding around a precisely formed core of a low-melting-point alloy.

Figure 8.13
Tennis racket made from reinforced plastics

The total moulding time of 3 minutes was a significant improvement over the 15 minutes required for the matched die moulding of thermosetting prepregs. Because of the labour-intensive assembly methods used for wooden rackets, the new production method was only 80% more expensive. The major advantages of injection moulding over prepreg moulding were:

- fewer manufacturing stages.
- shorter elapsed time for manufacture.
- reduction of work in progress.
- smaller manufacturing area.
- reduced labour content.

Testing

The rackets were evaluated initially in the laboratory and subsequently by players. A variety of rackets have subsequently been made by the process *(Figure 8.13)*.

Conclusion

The development was successful, not only in producing a cost-competitive product, but also in providing a strong base for further development. The technology was underpinned by a considerable investment in research and development over a period of two and a half years. The chief inference from this study is that a complete product redesign is often more likely to succeed than a replacement one. Such designs may well require new manufacturing techniques in order to make full use of the properties of reinforced plastics.

8.5 **Reference information**

General references

New designs are described in the current journals, which are listed in the reference section to Chapter 1. They are seldom of sufficient length, however, for the range of options to be discussed. Further examples are scattered through the proceedings of various conferences.

Specific references

8.1 **Golf club shaft**

'Sporting composites', in *Advanced Composites Engineering,* January 1990.

'Graphite golf club shafts, some questions answered', John Letters Ltd, Glasgow, 1990.

'Manufacturing considerations', discussion with Tri-Cast Composite Tubes Ltd, Rochdale, 1990.

8.2 **Road wheel**

'Structural automotive components in fibre reinforced plastics', AJ Wootton et al in *Composite Structures* 3, Elsevier, London, 1985.

The key performance requirements were obtained from *BS AU50 : Part 2, Section 5a : 1976, Appendices E and G*, and from *Guidelines for the testing of special wheels for cars*, covered by an operating licence granted in accordance with Section 22 of the *Strassenverkehrs-Zulassungs-Ordnung.*

'The drive is on for Britain's plastic car', D Bell in *Science and Business*, April 1988.

'Composite wheel cuts weight and machining', *Plastics and Rubber Weekly*, November 1989.

8.3 **Novel suspension**

'Development of a novel RP suspension system', A J Wootton in *Proc of 44th Annual Conference, Society of Plastic Industry*, 1989.

8.4 **Tennis racket**

'Some economic aspects of new materials', J D A Hughes and A L M McClintock in Materials Engineering Conference, Olympia, June 1987 (Promat DTI Initiative)

'Success stories', *Advanced Composites Engineering*, June 1989.

'Game, set and flash', D. Wise, *The Guardian* 16.6.92.

List of boxes

Accelerator: A material that, when mixed with a catalyst or resin, speeds up the curing process

Additive: A substance added to the resin to polymerize it like an accelerator, initiator, or catalyst or to improve resin properties such as filler or flame retardant

AFNOR: Association Français de Normalisation

ASME: American Society of Mechanical Engineers

ASTM: American Society for Testing Materials

Autoclave moulding: A process in which both heat and pressure are applied to a composite placed in an autoclave

BMC: Bulk moulding compound (see compounds)

BPF: British Plastics Federation

BS: British Standard

BSI: British Standards Institution

Catalyst: A material that when added in small quantities increases the rate of cure of a resin

CEC: Commission of the European Community

CEN: Comité Européen de Normalisation

Centrifugal moulding: A process in which chopped fibres, impregnated with resin, are sprayed into the inside of a mould which is rotated to permit consolidation of the mixture

CFM: Continuous filament mat

CFRP: Carbon fibre reinforced plastic

COM: Contact moulding fabrication process

Compound: An intimate mixture of a resin with other ingredients such as catalysts, fillers, pigments and fibres – usually contains all ingredients necessary for producing the finished product

Composite: A generic term to describe the mixture of a fibrous reinforcement and resin

Consolidation: A process in which the fibre/resin mixture is compressed to eliminate air bubbles and achieve a desired density

Contact moulding: A process in which the composite is laid up either by hand or by spraying short length fibres impregnated with resin inside a mould

CRP: Carbon fibre reinforced plastic

CSM: Chopped strand mat

Cure: A process in which a (thermosetting) resin becomes irreversibly hard – this involves a chemical reaction and may be accomplished using curing agents, with or without heat or pressure

Curing agents: Materials which help the cure process of a resin – these include accelerator, catalyst, hardener and initiator

Delamination: Separation between two or more reinforcement layers within a composite due either to incorrect processing or subsequent degradation during use

Denier: A numbering system for yarns and filaments in which the yarn number is equal to the mass in grams of 9000 m. The lower the denier, the finer the yarn

DIN: Deutsche Institut für Normung

DMC: Dough moulding compound (see Compound)

DNV: Det Norske Veritas, a classification society

DoE: Department of Environment, UK

Drape: The ability of a fabric to conform to a shape or surface

DTI: Department of Trade and Industry, UK

EEA: European Economic Area, the grouping of EC and EFTA countries

EC: European Community, a grouping (1992) of 12 member states

EFTA: European Free Trade Association

EN: European standard (* indicates draft or study)

ENV: A European pre-standard, ie a standard newly drafted which is circulated for a trial period

Fibre: A material in filamentary form having a small diameter compared with its length

Fibre content: The amount of fibre present in a composite, usually expressed as a volume percentage or weight fraction

Filament: See Fibre

Filament winding: A process which involves winding fibres or tapes onto a mandrel – these are generally pre-impregnated with resin

Filler: An inert material added to a resin to alter its properties or to lower cost or density – generally in the form of a fine powder

Finish: Material used to coat filament bundles – usually contains a coupling agent to improve the fibre to resin bond, a lubricant to prevent abrasion and a binder to preserve integrity of a filament bundle

Finished items: Made of fibre reinforced plastic, which are fully cured – such items generally consist of standard sections such as rod, bar, tube, channel or plate and are usually available ex stock from suppliers

Foam cores: A foamed resin created by using a foaming agent – rigid foams are useful as core materials in sandwich panels between stiffer outer layers

FRP: Fibre reinforced plastic

Gel coat: The surface layer of a moulding used to improve surface appearance or properties – applied using a quick-setting resin

Gel time: The time period from mixing of the curing agents with the resin until the mixture is sufficiently viscous that it does not readily flow

Geometric efficiency: Measure of the directionality of the fibre reinforcement

GMT: Glass mat thermoplastic

GRP: Glass reinforced plastic

Hardener: A substance added to a resin to promote curing

HM fibre: High modulus version of a reinforcing fibre

HS fibre: High strength version of a reinforcing fibre

Hybrid: A composite containing two (or more) types of reinforcement

IEC: International Electro-Technical Commission

IM: Injection moulding process

IMC: Injection moulding compound (see Compound)

IMO: International Maritime Organisation, London

Impregnation: The process of introducing resin into filament bundles or fabric laid up in a mould

Initiator: A substance which provides a source of chemical agents to promote curing

Injection: A process used for introducing a liquid resin (or heat-softened thermoplastic) into a mould

Injection moulding: A process which involves injecting a moulding compound into a mould

Interface: The area between the fibre and the matrix

ISO: International Standards Organization; also designates an International standard (* indicates draft or study)

JIS: Japanese Industrial Standard

Mandrel: A form of mould usually associated with a component having cylindrical symmetry

Mat: A material consisting of randomly orientated fibre bundles, which may be chopped or continuous, loosely held together with a binder or by needling

Matrix: The resin in which the fibrous reinforcement is embedded

Mould: The tooling in which the composite is placed to give the correct shape to the article once the resin has cured

Moulding: An article manufactured in a mould

Moulding compound: A type of compound suitable for moulding (see compound)

Moulding process: The process by which an article is made

NBS: National Bureau of Standards (USA)

Needle mat: A fabric which consists of short fibres felted together with a needle loom – a carrier may or may not be used

OCF: Owens Corning Fiberglas

PA: Polyamide resin (commonly known as nylon)

PAI: Polyamideimide resin

PAS: Polyarylsulphone resin

PEEK: Polyetheretherketone resin

PEI: Polyetherimide resin

PEK: Polyetherketone resin

PES: Polyethersulphone resin

Ply: A single layer in a laminate

PP: Polypropylene resin

PPM: Prepreg moulding

PPS: Polyphenylene sulphide resin

Preforming: A process by which fabric (or filaments) can be shaped into a desired form using a mould – this is achieved by coating the reinforcement with a small amount of thermoplastic binder

Prepreg: An intermediate fibre reinforced plastic product which is ready for manufacture into a component – either in the form of sheet (for moulding) or tape (for winding)

Prepreg moulding: A process in which prepreg material is moulded, either by autoclave or vacuum bag

Press moulding: A process in which a press (cold or heated) is used to form an article from a compound comprising a mixture of fibres and resin

PRM: Press moulding fabrication process

PUL: Pultrusion fabrication process

Pultrusion: A process in which filaments and/or fabric coated with resin, are pulled through a heated die and rapidly cured to retain the die shape

RIM: Reaction injection moulding process

RRIM: Reinforced reaction injection moulding

Reaction injection moulding: A process in which the resin and its curing agents are rapidly mixed and injected into a mould; reinforcing fibres may also be added and then designated reinforced reaction injection moulding (RRIM)

Reinforcement: Fibres which have desirable properties to increase the properties of the host matrix

Resin: The polymeric material in which the fibrous reinforcement is embedded

Resin transfer moulding: A process in which the catalysed resin is transferred into a mould into which the reinforcement has already been laid

Roving: A number of yarns, strands or tows which are collected into a parallel bundle with little or no twist

RTM: Resin transfer moulding process

Size: A treatment applied to yarns or fibres to protect their surface and facilitate handling

SMC: Sheet moulding compound

SOLAS: Safety of life at sea; regulations devised by IMO (qv)

Strand: An untwisted bundle of fibres

Swirl mat: A type of mat in which a continuous fibre bundle is randomly laid.

Tex: A unit of linear density equal to the mass in grams of 1000 metres of filament or yarn

Thermoplastic polymers: Plastics capable of being repeatedly softened by heat and hardened by cooling

Thermosetting polymers: Plastics that once cured by chemical reactions and/or heat, become infusible solids

Tow: A bundle of continuous filaments that are untwisted, typically 600 – 2,400 tex, may be either directly spooled or assembled bundles

TSC: Thermoplastic sheet compound

Twist: The number of spiral turns about the axis of a strand or yarn per unit length

UD: Unidirectionally aligned fibres possibly within a fabric

Vacuum bag moulding: A process in which a vacuum is used to consolidate prepreg material which is laid up in a mould and enclosed in a vacuum tight bag

Volume fraction: Fraction of fibres per unit volume of composite

Warp: Yarns running lengthwise in a fabric

Weft: Yarns running transversely in a fabric

Wet-out: Process by which the resin impregnates the yarns or fibres

WR: Woven roving – a plain coarse weave fabric usually made from glass fibre tows

XMC: Form of sheet moulding compound in which the fibre bundles are aligned

Yarn: A bundle of filaments that have been twisted – generally used for processing into fabrics, typically 68 tex

Note: This glossary is self-consistent with that proposed by the ASM International Handbook Committee as set out in Volume 2 of the Engineered Materials Handbook.